The Steel Market in 1997 and the Outlook for 1998 and 1999

ORGANISATION FOR ECONOMIC CO-OPERATION AND DEVELOPMENT

ORGANISATION FOR ECONOMIC CO-OPERATION AND DEVELOPMENT

Pursuant to Article 1 of the Convention signed in Paris on 14th December 1960, and which came into force on 30th September 1961, the Organisation for Economic Co-operation and Development (OECD) shall promote policies designed:

- to achieve the highest sustainable economic growth and employment and a rising standard of living in Member countries, while maintaining financial stability, and thus to contribute to the development of the world economy;
- to contribute to sound economic expansion in Member as well as non-member countries in the process of economic development; and
- to contribute to the expansion of world trade on a multilateral, non-discriminatory basis in accordance with international obligations.

The original Member countries of the OECD are Austria, Belgium, Canada, Denmark, France, Germany, Greece, Iceland, Ireland, Italy, Luxembourg, the Netherlands, Norway, Portugal, Spain, Sweden, Switzerland, Turkey, the United Kingdom and the United States. The following countries became Members subsequently through accession at the dates indicated hereafter: Japan (28th April 1964), Finland (28th January 1969), Australia (7th June 1971), New Zealand (29th May 1973), Mexico (18th May 1994), the Czech Republic (21st December 1995), Hungary (7th May 1996), Poland (22nd November 1996) and Korea (12th December 1996). The Commission of the European Communities takes part in the work of the OECD (Article 13 of the OECD Convention).

Publié en français sous le titre :
LE MARCHÉ DE L'ACIER EN 1997 ET LES PERSPECTIVES POUR 1998 ET 1999

FOREWORD

The OECD Steel Committee decided to undertake studies of the steel market and its outlook at its first meeting in 1978. Since that date, a report has been published yearly, beginning with *The Steel Market in 1978 and the Outlook for 1979*.

This report was prepared by Mr. Franco Mannato of the OECD Secretariat. The Steel Committee examined the report, which is based on data received before 31 March 1998. It is published on the responsibility of the Secretary-General of the OECD.

TABLE OF CONTENTS

INTRODUCTION

At its 50th meeting, in the spring of 1997, the Steel Committee decided, when adopting its programme of work, that a report on steel market trends in 1997 and the outlook for 1998 and 1999 would be drawn up in early 1998. After being discussed by the Steel Committee, the report will be published on the responsibility of the Secretary-General of the OECD, as in previous years.

Certain delegations to the Steel Committee provided statistics and other information on market developments in their countries, and the Secretariat has taken these into account. Since, however, it had to produce a coherent world outlook, it is possible that the text and the estimates may differ somewhat from those provided by the various delegations. It is therefore the Secretariat which is responsible for the forecasts.

Following the accession of the Czech Republic, Hungary, Korea and Poland as Member countries of the OECD during 1996, the statistics for these countries have been included in the OECD total and, as far as possible, in order to maintain a degree of coherence, the historical data have been recalculated on that basis. Also, since Brazil became a full participant in the Steel Committee in 1996, statistics for that country have been added to most of the tables and removed from those for the Latin American zone.

As a result of these changes, the data for the Czech Republic, Hungary and Poland have been removed from the central and eastern Europe zone and have been included in the "other Europe" zone. As regards the European Union, only the EU(15) zone remains and the historical data have been recalculated as far as possible.

As a result of the financial crisis in Asia, the Secretariat has split the "other Asia" zone into two parts: the "ASEAN(5)" area, which covers Indonesia, Malaysia, the Philippines, Singapore and Thailand, and the "rest of Asia", including North Korea, which is no longer grouped with China.

As far as possible, in addition to data for the New Independent States (ex-USSR) overall, the Secretariat has also tried to provide a breakdown for Russia, the Ukraine and the other NIS.

The report has been drawn up using the information received and the statistics available as at 31 March 1998.

2

NOTES ON THE MAIN FEATURES OF THE STEEL MARKET IN 1997, 1998 AND 1999

The main quantitative results for the steel market in 1997 and probable trends in 1998 and 1999 are contained in the statistical annex to the present document. The main developments in this market may be summarised as follows.

1997

Apparent steel consumption

- World: World steel consumption, which had levelled off in 1996 (–0.1 per cent), started to rise again in 1997, showing an increase of around 6.5 per cent on 1996. This increase reflects a fairly widespread trend, but was particularly noticeable in the OECD area as a whole and in Latin America, the Middle East and some of the other countries of Asia, while the ASEAN(5) countries saw their consumption decline.
- OECD. Following the 3 per cent fall in steel consumption in 1996, which was largely attributable to stock drawdowns, in 1997 consumption rose by 7.8 per cent, or nearly 30 million tonnes, in the OECD area as a whole. Totalling 414.5 million tonnes, in finished product equivalents, apparent steel consumption in the OECD area reached another record level in 1997.
- All Member countries reported increases in steel consumption. The largest were in Europe and North America, the smallest in Japan and Korea.
- Brazil, seems to have seen a strong recovery in apparent steel consumption – up over 19 per cent – in 1997 after the slight decline reported in 1996.
- As for other areas, in Latin American countries, demand for steel in 1997 was 7.7 per cent up on 1996. In South Africa, the increase was of the order of 6.3 per cent, whereas the rest of Africa reported a fall of around 3.4 per cent in steel consumption. In the Middle East, steel consumption increased for the tenth consecutive year in 1997, this time by around 8.5 per cent, and is now more than double the 1987 level. In India, steel consumption rose again for the eighth consecutive year, but at only 0.4 per cent the increase was substantially lower in 1997. In the five countries grouped under the heading ASEAN(5), consumption fell by slightly more than 8 per cent as a result of the financial crisis they experienced in the course of the year. In the other countries of Asia, apparent steel consumption rose by 12.9 per cent.
- The downward trend in steel consumption in the New Independent States (ex-USSR) finally reversed itself and in 1997 consumption rose by 9.7 per cent. The recovery was weakest in Russia, where consumption was up by only 1 per cent, while in the Ukraine and the other NIS, where the decline had halted in 1996, consumption was up by over 20 per cent in 1997.
- In the non-OECD countries of central and eastern Europe, apparent steel consumption rose by 3.3 per cent in 1997, but trends were very varied, with demand down 7.3 per cent in Romania, up 15.5 per cent in the Slovak Republic and up almost 49 per cent in Bulgaria.
- In China, demand for steel increased by 3.9 per cent, or 3.6 million tonnes more than in 1996.

◆ Graph I. **World apparent steel consumption**
Million tonnes product equivalent

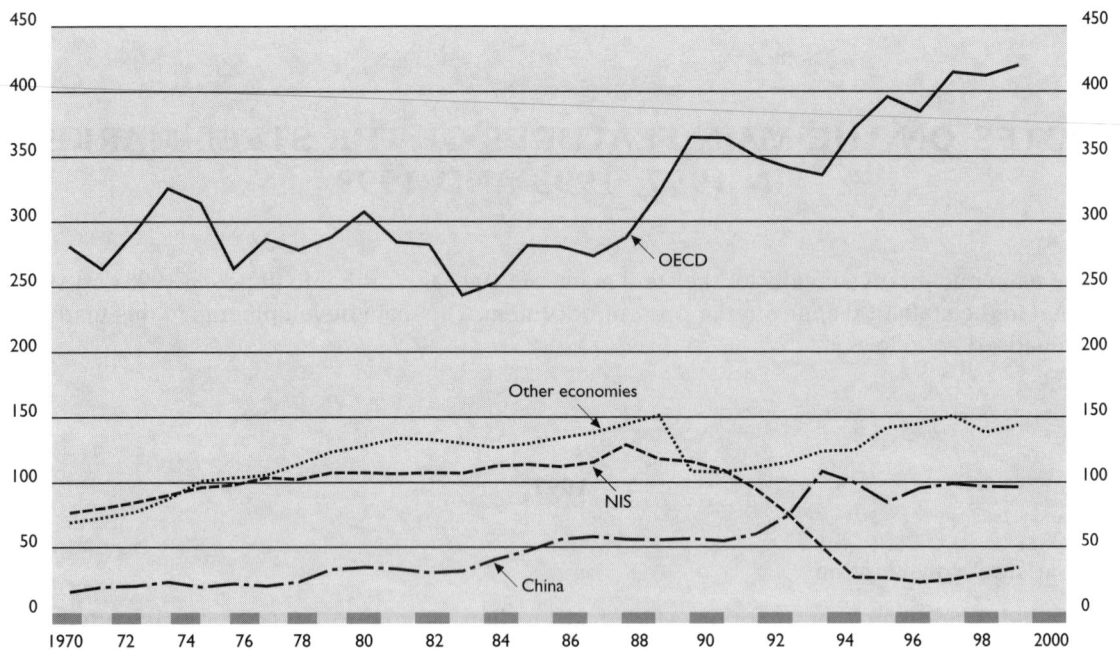

Source: OECD Secretariat.

Growth in apparent consumption of steel (AC) and estimated growth in real consumption of steel (RC) and total steel stocks held by steel producers consumers and merchants [1]

	OECD	Rest of OECD	Total for United States, EU and Japan			
					Total stocks of steel	
	AC	AC	AC	RC	Yearly change	End-year level
	In million tonnes of finished product equivalent					In weeks of real consumption
1986	274.8	41.1	233.7	239.6	−5.9	72.3	15.7
1987	289.0	43.1	245.9	243.9	+2.0	74.3	15.9
1988	324.6	45.9	278.7	261.1	+17.6	91.9	18.3
1989	330.4	46.2	284.2	279.9	−4.3	96.2	17.9
1990	330.8	40.3	290.5	286.8	+3.7	99.9	18.1
1991	316.7	38.5	278.2	281.9	−3.7	96.2	17.7
1992	310.9	40.7	270.2	272.3	−2.1	94.1	18.0
1993	306.0	44.4	261.6	264.2	−2.6	91.5	18.0
1994	334.2	49.5	284.7	284.7	0.0	91.5	16.7
1995	395.6	88.4	307.2	301.6	+5.6	97.1	16.7
1996	384.7	90.0	294.7	301.0	−6.3	90.8	15.6
1997	414.5	99.3	315.2	310.4	+4.8	95.6	16.0
1998e	411.9	93.2	318.7	318.3	+0.4	96.0	15.7
1999f	419.8	96.6	323.2	323.7	−0.5	95.5	15.3

e: Estimate.
f: Forecast.
1. In previous years, figures for apparent steel consumption can be derived, as they have in this report, from the data available on steel production and trade. Variations in apparent consumption are due to variation in real consumption and/or changes in the total steel inventories maintained by steel producers, consumers and merchants. Data regarding the level of, or annual variations in, both these parameters, however, are far from complete. The figures given for real consumption and annual variations in total stock levels should therefore be taken as "reasonable" estimates of two inter-related factors. Furthermore, in calculating the level of total steel stocks in tonnage terms by the end of 1984, it has been assumed that the stocks were equal to 18 weeks for estimated real consumption for that year (*i.e.* 8 weeks for producers and 10 weeks for consumers and merchants). For the years after 1984, the level of total steel stocks was first calculated in terms of tonnage, based on the estimated annual variation, and subsequently expressed in terms of weeks of real consumption.
2. As from 1995, data concerning the EU are related to EU (15) and the OECD total includes the new countries: the Czech Republic, Hungary, Korea and Poland.
Source: OECD Secretariat.

– In the OECD area steel stocks, which had fallen substantially in 1996, particularly in Europe, increased overall in 1997. As a result of this trend in steel stocks, real steel consumption in the OECD area rose by only a little over 3 per cent in 1997.

Steel trade

– World trade in steel (excluding intra-EU trade) started to pick up again in 1997, by 2.5 per cent (by volume) compared with 1996. World trade in steel accounted for nearly 26 per cent of world steel consumption in 1997.

– Steel exports from the OECD area rose by 3.6 per cent in 1997, or 3.3 million tonnes more than in 1996. As a result of higher consumption, steel imports rose by 5.5 million tonnes, or 7.4 per cent. Consequently net OECD exports were down by 11.9 per cent.

– Steel imports in the United States were up 7.1 per cent, to 28.9 million tonnes, in 1997 as a result of strong demand and the rise of the dollar against the currencies of other steel-producing countries. Exports of steel products also increased by 20.3 per cent, but these were exclusively to North America. Despite a healthy domestic market, the share of imports on the American market rose from 26.3 per cent in 1996 to 26.8 per cent in 1997.

– As a result of the strong upturn in the demand for steel in the European Union, imports rose by around 6.6 per cent while exports declined by 6.8 per cent.

– In Japan, steel imports, which had fallen by 15.6 per cent in 1996, went up by 8 per cent in 1997. At the same time, steel exports rose by 18.6 per cent, bringing the country's net exports up to 23.3 per cent However, the share of imports on the Japanese market rose from 7.5 per cent in 1996 to 7.9 per cent in 1997.

– Net imports increased by 50 per cent in Latin America, 10.3 per cent in India and 9.5 per cent in the Middle East, while net exports remained unchanged in South Africa. Net imports in the ASEAN(5) group fell by 10.3 per cent and by 0.6 per cent in the other countries of Asia.

– In China, a 5.1 per cent fall in imports and a rise of 13 per cent in exports combined to produce a decline in net imports of 16.9 per cent.

– Net steel exports from the NIS overall stayed at 1996 levels, although Russia's were down slightly, while net exports from the Ukraine and the other NIS continued to show a slight rise.

Crude steel production

– World: World crude steel production increased by 6.2 per cent in 1997 to 794.1 million tonnes, a new record level.

– OECD: Crude steel production for the area as a whole rose by 6.7 per cent in 1997, or 30 million tonnes more than in 1996. Output totalled 481.2 million tonnes. Production increased in all Member countries in the area. In Brazil, too, crude steel production increased by 3.6 per cent.

– Crude steel production declined again in Africa by 20.3 per cent, but increased by 3.9 per cent in South Africa, 5.7 per cent in Latin America and 7 per cent in the Middle East. In India, production remained at the same level as in 1996, *i.e.* 23.8 million tonnes. In the ASEAN(5) group, production began to decline – down 2.7 per cent – but in the other countries of Asia it rose by a substantial 29.5 per cent.

– In the NIS overall crude steel production rose by 3.5 per cent. In Russia, production again declined by 1.5 per cent, but rose by 6 per cent in the Ukraine and 17.7 per cent in the other NIS.

– In the countries of central and eastern Europe, steel production rose by 7.8 per cent.

– In China, crude steel production reached another new record of 107.6 million tonnes, an increase of 6.3 per cent on 1996. For the second year running, China was the world's leading steel producer.

Steel capacity utilisation rate

– In the OECD area overall, the average capacity utilisation rate increased to 82 per cent in 1997.

– Capacity utilisation rates exceeded 90 per cent in Australia, Canada, Korea and Mexico. Utilisation rates were 89 per cent in the United States, 81 per cent in the European Union and 70 per cent in Japan.

– In most other areas of the world, with the exception of the NIS and the countries of central and eastern Europe, steel production capacity continued to increase. In China, capacity utilisation exceeded 88 per cent, while world-wide the average capacity utilisation rate in 1997 was just over 75 per cent.

Steel prices

– The upturn in demand for steel in the OECD area and the build-up of stocks prompted a rise in steel product prices in the course of 1997. The rise was relatively low – less than 1 per cent over the year – in the United States, but was higher in Europe, where prices increased throughout the year, especially for flat products.

1998

Apparent steel consumption

– World: The financial crisis that hit certain Asian countries in 1997 is likely to have an impact on world steel consumption in 1998. Initial estimates indicate that apparent steel consumption world-wide could fall by just under 2 per cent in 1998 compared with 1997, *i.e.* by some 15 million tonnes.

– OECD: The OECD area is not expected to escape the world-wide decline and steel consumption in the area overall could fall by just over 0.5 per cent, *i.e.* by 2.5 million tonnes on 1997 levels. However, trends in the different Member countries will be very mixed.

– While steel consumption is expected to remain very steady in the United States, it is likely to pick up in Australia and particularly in the European Union, where the increase could be as much as 4.8 per cent. Steel consumption is expected to be flat in Mexico and decline, though remaining at high levels, in Canada. It is in Japan, particularly, and more so in Korea, that steel consumption is expected to fall in 1998.

– Demand for steel looks set to continue rising in the rest of South America, in South Africa, and to a lesser extent in the Middle East. Consumption in Africa may even pick up. Conversely, in India, the ASEAN(5) group and the other countries of Asia, consumption is expected to decline by amounts ranging from 5 per cent in India to 26 per cent in the ASEAN area.

– In the countries of central and eastern Europe, steel consumption is expected to continue to rise.

– The recovery in demand for steel, which finally seems to have occurred in the NIS in 1997, should continue at a faster rate in 1998, when demand should be up by over 10 per cent.

– Steel demand in China should remain at much the same level as in previous years, *i.e.* over 95 million tonnes, but could well decrease by a little over 2 per cent compared with 1997.

Steel trade

– In volume terms, world trade in steel is expected to be 7.5 to 8 per cent down on 1997, a fall of almost 13 million tonnes, largely due to the effects of the crisis in Asia.

– Total net exports from countries in the OECD area are likely to fall by around 6.9 per cent, reflecting a decline in world trade, with imports expected to fall slightly less than exports.

– Net exports from the European Union are expected to fall as a result of increasing demand on the internal market accompanied by rising imports and falling exports. Net exports are also expected to decline in Japan, but this time due to a decline in both exports and imports caused by a marked downturn on its domestic market.

– In the United States, net steel imports should fall by about 10 per cent, since the additional capacity installed in 1997 should reduce imports, bringing their share of the American market down from 26.8 per cent in 1997 to 25.6 per cent in 1998. Korea's net exports could more than double as a result of a very sharp decline in imports combined with a substantial increase in exports.

– In Latin America and the Middle East, net steel imports are expected to increase, but should decline in India. It is mainly in the other countries of Asia that net steel imports are expected to show the most substantial decline, falling by 30 to 40 per cent.

– Steel exports from the NIS, a substantial proportion of which have been to South-East Asia, should decline quite substantially, by around 5 million tonnes, or 12 per cent on 1997 levels. It may be that this decline will be accompanied by the diversion of a substantial share of the remaining exports.

– China's net imports are expected to increase appreciably, as a result of a steep fall in exports, particularly to the other countries of Asia, and an increase in steel imports.

Crude steel production

– World: As a result of the slowdown in world demand, crude steel production should also decline by a little over 2 per cent in 1998, *i.e.* by some 17 million tonnes on 1997 levels.

– OECD: Crude steel production in the area as a whole should fall by just under 1 per cent under the combined impact of lower production in Korea and Japan. In other Member countries, the trend towards an upturn in crude steel production should continue, but at a much slower rate than in 1997. Crude steel production in Brazil could also increase by around 1 per cent.

– While crude steel production in South America and South Africa could increase further in 1998, it should fall in practically all of the other regions, particularly in the ASEAN(5) group, where the decline is expected to be at least 16 per cent.

– In the non-OECD members among the central and eastern European countries, crude steel production should continue to rise, again at a slower rate than in the previous year.

– As for the NIS, the substantial fall in exports should lead to a decline in production, particularly in Russia.

– China could also see a substantial decline in crude steel production. However, it is expected to implement a vast restructuring and reorganisation plan for its steel industry.

Crude steel production capacity utilisation

– In 1998, crude steel production capacity continued to increase, by around 3 per cent world-wide. However, it is already clear that a number of projects planned for the future have either been cancelled or postponed as a result of the Asian crisis.

– In the OECD area, production capacity in 1998 is expected to increase by 1.5 per cent on 1997 levels, but the fall in steel production should bring the average capacity utilisation rate down to 80 per cent.

– In China, production capacity should increase by a further 5.4 per cent or 6.5 million tonnes per year, but the capacity utilisation rate should fall from 89 per cent in 1997 to 77 per cent in 1998. In the other regions, particularly in Asia, the capacity utilisation rate should fall substantially, possibly to under 50 per cent.

Steel prices

– While steel prices continued on an upward trend during the first quarter of 1998, falling demand in Asia and the possible diversion to other destinations of substantial volumes of steel exports will probably lead to pressure on prices in the second half of the year.

1999

Apparent steel consumption

– World: Following the decline in world demand for steel expected in 1998, we may see apparent consumption recover slightly – by as much as 2.5 per cent – in 1999.

– OECD: With the recently revised figure of 2.5 per cent for economic growth for the area as a whole in 1999 – growth which should be led by private consumption and investment – apparent steel consumption in the OECD area could begin to recover, increasing by just under 2 per cent in 1999.

– Steel consumption should continue to increase in Europe, and could show signs of recovery in Korea, Japan and Mexico. In Australia, Canada and the United States, however, it could show a slight decline, although remaining at high levels.

– In Brazil, demand for steel should continue to increase substantially – by over 6 per cent – in 1999.

– While steel demand should continue rising in Latin America, in Africa and in the Middle East, in India it could show signs of a modest recovery. In the other countries of Asia, after the steep declines reported by some countries in 1998 and 1997, steel demand could also begin to pick up.

– In the NIS, the recovery in steel demand noted since 1997 should speed up, especially in Russia. However, it should be noted that the increase of 15 per cent in steel consumption forecast for the NIS overall in 1999 will only bring consumption to one-quarter of the record set by the USSR in 1987.

– In China, demand for steel in 1999 could remain at much the same level as in 1998, levelling off before increasing in subsequent years.

Steel trade

– World trade in steel could again fall slightly, by about 3 per cent, accounting for only 22.7 per cent of the world steel consumption. This is the likely outcome of commissioning additional capacity in different parts of the world.

– Total net exports from the OECD area should again be higher than in 1998, but the increase would be due chiefly to a far more substantial decline in imports than in exports. While net exports from Japan are expected to increase, those from the EU(15) will continue declining as a result of the continuing growth in internal demand. In contrast, net imports by the United States are expected to decline again substantially, bringing the market share of imports to less than 22 per cent.

◆   Graph 2.   **OECD apparent steel consumption, 1970-99**
Million tonnes product equivalent

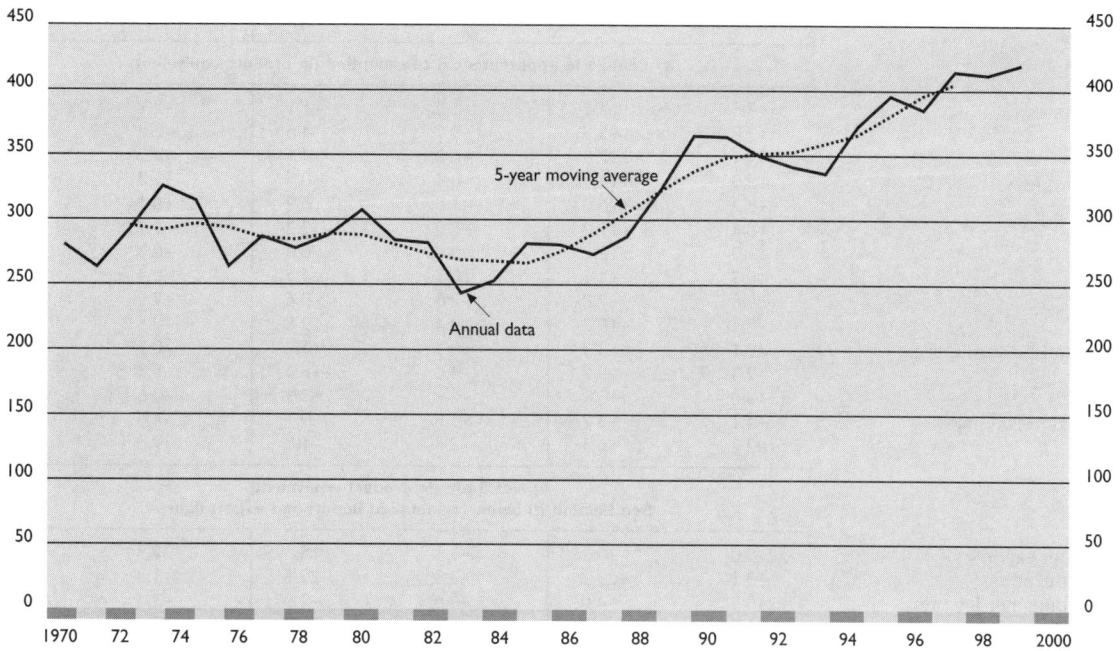

Source: OECD Secretariat.

◆   Graph 3.   **OECD crude steel capacity and production, 1970-99**
Indices, 1973 = 100

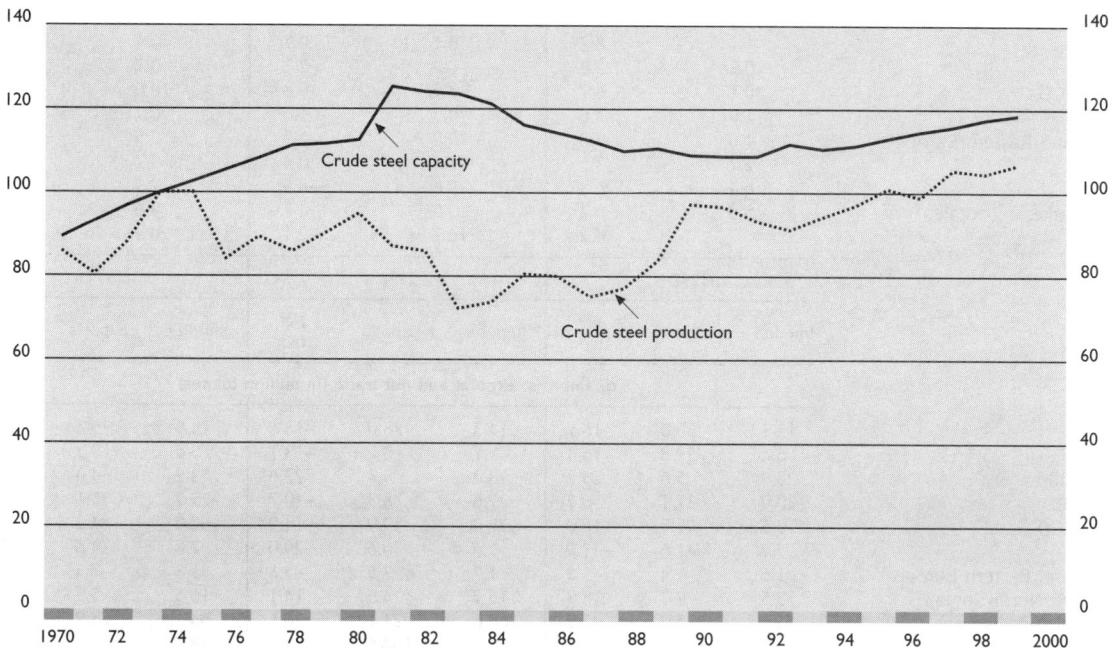

Source: OECD Secretariat.

Main results

	1997 variation in		1998 variation in		1999 variation in	
	Million tonnes	%	Million tonnes	%	Million tonnes	%
a) Change in apparent steel consumption (in product equivalent)						
United States	+3.6	+3.5	+1.2	+1.1	−2.1	−1.9
Japan	+2.2	+2.9	−3.7	−4.6	+2.4	+3.1
EU (15)	+14.7	+12.8	+6.0	+4.8	+4.2	+3.1
Other Europe	+3.5	+14.7	−0.4	−0.5	+2.3	+8.6
Canada	+3.2	+26.0	−1.0	+6.2	+0.1	−0.7
Korea	+0.4	+1.0	−5.0	−13.2	+1.1	+3.3
Mexico	+1.9	+18.2	0.0	0.0	+0.3	+2.3
Oceania	+0.3	+4.1	+0.3	+4.4	−0.1	−0.6
Total OECD	+29.8	+7.8	−2.6	−0.6	+7.9	+1.9
Brazil	+2.3	+19.2	+0.4	+2.8	+0.8	+6.3
Central and Eastern Europe	+0.1	+3.3	+0.3	+5.9	+0.7	+13.4
NIS	+2.3	+9.7	+4.6	+18.4	+4.3	+14.5
China	+3.6	+3.9	−2.2	−2.3	−0.6	−0.6
Rest market economies	+4.1	+3.2	−12.8	−9.7	+4.6	+3.9
World	+42.2	+6.5	−12.3	−1.8	+17.8	+2.6
b) Net trade (in product equivalent)						
See Section d) below for detailed import and export figures						
EU (15) net exports	−3.0	−15.2	−3.1	−18.6	−2.3	−16.9
Japan net exports	+3.1	+23.3	−3.3	−20.1	+1.2	+9.2
United States net imports	+0.9	+4.0	−0.6	−2.6	−3.4	−15.0
Rest OECD net exports	−1.3	−4.7	−0.9	−17.6	77.0	−8.3
Total OECD net exports	−2.1	−11.6	−1.1	−6.9	+1.4	+9.4
China net imports	−1.6	−16.2	+5.8	+69.9	−3.1	−22.0
Other economies net imports	−1.9	−4.9	−2.5	−19.8	−63.6	89.3
c) Crude steel production						
United States	2.8	3.0	2.0	2.0	1.5	1.5
Japan	5.8	5.8	−7.6	−7.2	3.9	4.0
EU (15)	13.0	8.8	3.5	2.1	1.7	1.1
Other Europe	2.8	8.2	0.2	0.5	1.7	4.6
Canada	0.8	4.9	0.2	1.6	0.0	0.0
Korea	3.7	9.4	−2.4	−5.5	0.8	2.0
Mexico	1.1	8.2	0.1	0.7	0.4	3.2
Oceania	0.3	2.8	0.2	2.2	−0.1	−1.5
Total OECD	30.1	6.7	−3.7	−0.8	10.0	2.1
Brazil	1.0	3.6	0.2	1.0	0.7	2.4
Central and Eastern Europe	1.0	7.8	0.3	2.3	0.5	4.0
NIS	2.7	3.5	−0.4	−0.4	3.2	4.0
China	6.4	6.3	−9.6	−8.9	2.9	3.0
Other market economies	5.0	6.2	−3.5	−4.1	3.6	4.4
World	46.2	6.2	−16.7	−2.1	20.9	2.7

	1997			1998			1999		
	Imports	Exports	Net trade	Imports	Exports	Net trade	Imports	Exports	Net trade
d) Imports, exports and net trade (in million tonnes)									
EU (15)	13.3	30.0	−16.7	14.3	28.0	−13.6	14.5	25.8	−11.3
Japan	6.4	22.8	−16.4	5.1	18.2	−13.1	5.0	19.3	−14.3
United States	28.9	5.6	23.2	28.0	5.4	22.6	23.2	4.0	19.2
Rest OECD	30.9	37.1	−6.2	27.6	38.3	−10.7	25.2	35.1	−9.9
Total OECD	79.5	95.5	−16.0	75.0	89.9	−14.9	67.9	84.2	−16.3
NIS	7.7	41.6	−33.9	7.6	36.6	−29.0	7.8	34.8	−27.0
Central and Eastern Europe	1.6	7.4	−5.8	1.7	7.5	−5.8	1.8	7.3	−5.5
China and North Korea	15.5	7.3	8.3	18.7	4.6	14.1	16.5	5.5	11.0
Rest market economies	72.5	25.2	47.3	60.6	24.1	36.5	64.1	25.9	38.2
World	176.8	177.0	−0.2	163.6	162.6	1.0	158.1	157.7	0.4

Main results *(cont.)*

	Capacity in million tonnes			Utilisation rate in %		
	1997	1998	1999	1997	1998	1999
	e) Crude steel production capacity[1] and utilisation rates					
United States	109.3	113.3	113.3	89.2	87.8	89.1
Canada	16.7	17.0	17.0	92.8	92.4	92.4
Korea	45.8	49.2	54.2	93.0	81.7	75.6
Mexico	15.3	16.5	17.0	93.5	87.3	87.1
Oceania	9.8	9.8	9.8	96.9	99.0	98.0
Japan	149.5	149.5	149.5	70.0	64.9	67.5
EU (15)	196.6	197.4	195.1	81.3	82.8	84.6
Other Europe	43.4	44.4	44.7	86.4	84.9	88.1
Total OECD	586.4	595.1	600.6	82.1	80.2	81.2
Brazil	30.0	30.0	32.0	87.2	88.0	84.5
NIS	141.9	141.9	143.6	56.2	56.0	57.5
Central and Eastern Europe	16.4	15.7	16.0	79.9	85.4	86.9
China	121.2	127.7	128.5	88.8	76.7	78.5
Other economies	149.2	166.7	189.8	57.8	49.6	45.5
World	1 048.1	1 080.9	1 113.5	75.8	71.9	71.7

1. Estimates of capacity for OECD countries are based on the Steel Committee's Annual Survey of Effective Capacity. There are differences in country definitions, and the changes in a country's operating rates from year to year are more significant than direct comparisons between the various countries' operating rates.
Source: OECD Secretariat.

– Steel imports should continue to rise in Latin America and the Middle East and should begin to pick up in the countries of Asia. In contrast, in India a sharp fall in imports is expected along with an increase in exports, which will make India a net exporter.

– Net exports of steel from the NIS are expected to decline again, as are those from the countries of Central and Eastern Europe.

– In China, steel imports should decline, while exports are expected to pick up again.

Crude steel production

– World: As a result of the expected recovery in steel consumption, crude steel production should also increase, by just over 2.5 per cent, in 1999.

– OECD: Steel production in the area as a whole could increase by about 2 per cent, or 10 million tonnes more than in 1998.

– Among Member countries generally, production is expected to increase, except in Australia and Canada where it should remain close to 1998 levels.

– In Brazil, crude steel production should again increase by around 2.5 per cent, exceeding 27 million tonnes.

– Crude steel production should also remain on an upward trend in all other regions and in the NIS and China, where it is expected to pick up again.

Steel production capacity utilisation

– Crude steel production capacity in the OECD area as a whole is expected to increase by a further 5.5 million tonnes in 1999, to over 600 million tonnes per year or 54 per cent of world capacity. The average capacity utilisation rate should be in the region of 81 per cent.

– In all of the other areas, steel production capacities will continue to increase but at a slower rate, which should lead to widespread improvements in utilisation rates provided that the expected recovery in output materialises.

Steel prices

– As forecast, steel market trends in 1999 should result in higher prices for steel products, which will be more noticeable from the second quarter onward.

3

DEVELOPMENTS IN THE STEEL MARKET BY AREA

UNITED STATES

Economic growth in the United States remained very strong in 1997. GDP rose by 3.8 per cent over the year, the strongest in the last nine years and as high as 3.9 per cent, at annual rates, in the fourth quarter. Private consumer spending, which accounts for some two-thirds of GDP, increased by 3.3 per cent over the year, reflecting the high level of consumer confidence in the US economy and an increase in disposable income. Investment, up 7.7 per cent on 1996, also continued to boost the economy. Over the whole of 1997, 3.3 million jobs were created and the average unemployment rate was 5 per cent. Despite strong growth and low unemployment rates, inflation remained under control. Wholesale prices were down 1.2 per cent and consumer prices rose by only 1.7 per cent over the year.

Capital investment jumped by 12.2 per cent. Output in the automobile manufacturing sector rose to 11.7 million units, up 2.2 per cent on the already very high 1996 levels. Activity in the construction sector was up by 2.2 per cent on 1996. New starts in the residential construction sector were virtually unchanged, but new house sales were up by 5.7 per cent in 1997 to 800 000, the highest level since 1978, which is a good indication that consumer confidence is high.

In 1998, economic growth should slow slightly and the GDP growth rate could be around 2.7 per cent. This slowdown will reflect the adverse impact of the Asian crisis on the American economy. Indeed, the crisis can be expected to widen the United States trade deficit with, on the one hand, the collapse of major foreign markets and, on the other, low-priced imports taking a larger share of the market. At this point, it is still difficult to gauge exactly what impact the Asian crisis will have on the economy of the United States, but it is generally expected that the GDP growth rate will be 0.5 to 1 per cent down in 1998.

Private consumer spending could continue to rise, at an estimated rate of 3.8 per cent. Likewise, investment could be 8.2 per cent up on 1997. The unemployment rate should continue to decline, to around 4.8 per cent. Industrial output could again rise by approximately to 4.9 per cent. Automobile production is not expected to show an increase on 1997, but will remain high. In the construction sector, residential construction is expected to increase by 2 to 3 per cent.

For 1999, the forecasts currently available suggest that the growth of the US economy could revert to its long-term trend, showing an increase of around 2.1 per cent. Private consumer spending should

Production indices, 1990 = 100 (seasonally adjusted)

	1995	1996	1997	1997 fourth quarter	1997/96 % change
Industrial production	116	120	126	129	5.0
Manufacturing industries	118	122	129	132	5.7
Motor vehicles and parts	138	137	144	147	5.1
Fabricated metal products	115	118	121	124	2.5
Non-electrical machinery	143	155	171	175	10.3
Electrical machinery	195	227	264	283	16.3
Mining	99	100	102	103	2.0

Source: OECD, *Indicators of Industrial Activities.*

again increase by 3 per cent, with investment increasing by a further 4.9 per cent. The US trade deficit could continue to increase. The growth in industrial output should slow to no more than 0.7 per cent.

Benefiting from the continued economic growth in 1997, the steel market again registered record levels of demand. Apparent steel consumption, expressed as finished product equivalents, reached a record 107.4 million tonnes, up 3.5 per cent on 1996. American steelmakers' deliveries to the domestic market rose by 4.6 per cent to 95.7 million tonnes, the highest level since 1974. Deliveries to the construction sector were up by just over 10 per cent, to the processing sector by 12.6 per cent and to the agricultural sector by a massive 18 per cent. Deliveries to the key oil sector rose by 5.6 per cent. Conversely, deliveries to the automobile sector and dealers declined by 0.4 per cent and 2.2 per cent respectively.

By product category, pipe performed excellently, with deliveries up 11.1 per cent, mainly to the energy industry. Demand for strip products also rose, with deliveries up by 10.9 per cent, thanks to strong demand from the construction sector. Deliveries of flat products, which account for almost half of all deliveries, rose by only 3.1 per cent, as a result of strong competition from imports.

Crude steel production increased by 3 per cent, to 97.5 million tonnes. However, a 3.9 million tonne increase in crude steelmaking capacity led to a slight decline in the capacity utilisation rate to 89.2 per cent in 1997 compared to 90 per cent in 1996. Continuously cast steel production increased by 3.7 per cent in 1997 and now accounts for 94.7 per cent of all the steel produced in the United States. Steel product prices improved slightly in 1997, up 0.7 per cent on the previous year, but without reaching 1995 levels. Nonetheless, this slight increase, along with productivity gains, substantially boosted the industry's performance.

As regards trade in steel products, with high domestic demand and the rise of the dollar against the currencies of other steel-producing countries steel imports rose by 7.1 per cent to a record 28.9 million tonnes. The rate of import penetration on the American market went up from 26.3 per cent in 1996 to 26.8 in 1997. Imports from Russia, which again strengthened its position as a supplier to the American market in 1997, doubled to 3 million tonnes. The stronger dollar and the financial crisis in Asia also led to an increase in imports from that region. Imports from Japan and Korea rose by 27.2 per cent and 16.4 per cent respectively and imports from Chinese Taipei by 31.5 per cent. Some Latin American countries also increased their exports to the United States, while imports from the EU(15) fell by 15 per cent and imports from Canada were down 2.8 per cent.

Surprisingly, the trend in steel exports from the United States was positive, up 20 per cent to 5.6 million tonnes in 1997. However the increase was almost exclusively in exports to North America, which accounted for 79 per cent of US exports in 1997 as compared with only 69 per cent in 1996. Exports to South-East Asia, which seemed so promising a few years ago, virtually dried up. In fact, exports to the four main markets in the area, China, Chinese Taipei, Japan and Korea fell by 80 per cent and more compared with 1996.

The economic growth again forecast for the United States in 1998, and the favourable outlook for the main steel-consuming industries, buoyed up by a high level of investment, should see another if more modest increase in consumption of just over 1 per cent. A large increase in capacity, of at least 4 million tonnes per year mainly for flat products in mini-mills, and an extremely competitive environment, should lead to a reduction in imports. The Secretariat estimates that this reduction will be of the order of 3 per cent, but it could be as high as 5 per cent or more. The increase in production capacity could encourage producers to try to increase their exports, however low export prices and the effects of the Asian crisis suggest rather a decline in exports. Crude steel production could still increase by approximately 2 per cent in 1998, to around 99.5 million tonnes, The rate of capacity utilisation is expected to decline to around 87.8 per cent.

As for 1999, the steel market in the United States could flag, with apparent consumption declining by a little under 2 per cent. This said, crude steel production should still continue to increase, by around 1.5 per cent, to 101 million tonnes. Net steel imports by the United States are expected to fall by around 15 per cent, reflecting a decline in imports of 17 per cent or more and a fall in exports of just over 25 per cent. The rate of penetration of imports on the American market could fall to under 22 per cent, whereas the capacity utilisation rate should increase again to over 89 per cent.

CANADA

Economic growth in Canada picked up again in 1997, on the strength of business and household spending, boosted by low interest rates. The GDP growth rate stood at 3.8 per cent in 1997, compared with 1.5 per cent in 1996. Inflation was low, at around 1.6 per cent, as in 1996. The unemployment rate fell to an average of only 9.2 per cent in 1997 compared with 9.7 per cent in 1996.

Industrial output was up 3.9 per cent. The increase in disposable income and improved consumer confidence together with low interest rates were good for many sectors of industry. In the automobile sector, vehicle production in 1997 increased by 10.4 per cent to 2.4 million units, the highest ever in Canada. At the same time, new vehicle sales rose by 18.3 per cent to 1.4 million units. The continuing growth in demand for vehicles in Canada is also attributable to the fact that more than 5 million vehicles on the road are over 10 years old. In the construction sector overall, activity was up by 19.4 per cent: by 23.7 per cent in the non-residential sector and 16.5 per cent in the residential sector. Activity in the pipe product sector was up 14.2 per cent in 1997.

Crude steel production in Canada increased by 1.6 per cent in 1997 to 15.5 million tonnes, the highest level since 1980, of which 98.2 per cent was continuously cast steel. Total steel imports were up by more than 40 per cent on 1996, to 6.63 million tonnes. While, imports of semi-finished products declined by 22 per cent, imports of other steel products rose by over 60 per cent. The rate of import penetration in the Canadian market climbed to 37 per cent in 1997, compared with 27.9 per cent in 1996. At the same time, steel exports fell by 12.5 per cent, to 4.2 million tonnes. Apparent steel consumption, expressed as finished product equivalents, rose by 26 per cent in 1997. Prices of steel products in Canada firmed up a little in 1997 and, combined with the increase in output, enabled Canada's steel industry to increase its earnings by just over 8 per cent.

Projections for the Canadian economy in 1998 and 1999 forecast steady growth, and GDP could be up by around 3.3 per cent in 1998 and 3 per cent in 1999. Inflation should remain under control, although it could creep back up to 1.8 per cent in 1998 and 2 per cent in 1999. The economic trends should also enable a reduction in the rate of unemployment to only 8.7 per cent in 1998 and 8.3 per cent in 1999. Investment should continue to rise steadily at over 6.5 per cent and private consumption should also continue to rise steadily.

Industrial output is likely to continue increasing in 1998 and 1999 at the same rate as in 1997, *i.e.* 3.9 per cent. Production in the automobile sector should remain strong, given that demand for replacement vehicles is expected to be high until the year 2000. In the construction sector the upturn registered since 1994 in the non-residential sector is expected to continue as it is in the residential sector, on the strength of relatively low interest rates, at over 8 per cent in 1998.

However, apparent steel consumption is likely to fall in 1998 and 1999, although chiefly as a result of a substantial drop in imports, particularly of semi-finished products, as new steel capacity will have been commissioned. In 1998 steel exports may remain at 1997 levels, declining slightly in 1999. Crude steel production is likely to continue increasing, by 1.6 per cent in 1998, remaining at around 15.7 million tonnes in 1999.

MEXICO

The Mexican economy expanded vigorously in the first half of 1997, during which period GDP grew in real terms at a seasonally adjusted annual rate of 8 per cent. Exports and investment continued to rise rapidly, and private consumption began to pick up. Growth in GDP over the year as a whole amounted to 7 per cent. Investment in the construction sector was boosted by new infrastructure projects. Renewed growth in total domestic demand helped to fuel increased imports. Although the improvements in the labour market were not consistent, for the first time since the 1994/95 crisis wage increases were in line with, or even higher than, the rate of inflation. Nominal interest rates continued to fall until October 1997. The current direction of government policy, namely its firm commitment to continued disinflation and new public investment projects, should ensure sustained growth in the longer term. The most recent forecasts for 1998 indicate that GDP should grow by around 5.3 per cent, investment should increase by over 15 per cent and private consumption by 4.5 per cent. The economic climate should continue to remain favourable in 1999, with GDP growth of around 4.9 per cent, a 14 per cent increase in investment and 4.2 per cent growth in private consumption.

After the strong recovery in apparent steel consumption observed in 1996 (+27.5 per cent), demand for steel rose by over 18 per cent in 1997 to a new record level of 12.1 million tonnes. To meet this demand, steel imports rose by around 26 per cent and at the same time exports fell dramatically by 46.8 per cent to no more than 0.84 million tonnes. Crude steel production rose by 8.2 per cent to 14.3 million tonnes, another new record. The rate of utilisation of crude steel production capacity amounted to 93 per cent.

Steel demand is expected to remain steady in 1998 at the very high level reached in 1997. Steel exports should continue to decline by around a further 4.8 per cent, while imports are expected to fall by some 11 per cent. The upward trend in crude steel production should continue, but the actual increase is not expected to amount to more than 0.7 per cent. In 1999, steel demand may well start to rise again, although only modest growth of around 2.3 per cent is expected, which means that steel imports will probably increase too, by slightly over 5 per cent, particularly if exports also start to rise. Crude steel production could therefore rise by 3.2 per cent to 14.8 million tonnes. As new production capacity starts to come on stream (about 1.7 mtpa between 1997 and 1999), the average utilisation rate of production capacity in 1999 may well fall to 87 per cent.

EUROPEAN UNION (15)

The economic performance of the EU(15) area in 1997 proved to be better than expected with GDP growth of around 2.6 per cent, fuelled partly by exports. Growth nonetheless remained below average in Austria, Belgium, Italy and Sweden. Private consumption rose by 2.1 per cent, buoyed by wage increases, albeit modest, in a number of countries, and increased employment. Investment started to pick up again and rose by 2.4 per cent. The average rate of inflation again declined and in 1997 amounted to no more than 2.1 per cent.

This increase in economic activity in 1997 helped to fuel strong growth in the manufacturing sector in general in response to increases in both domestic demand and demand for exports. Industrial output rose by an average of 3.1 per cent in the EU(15) area, although growth was higher in Spain (+6.8 per cent) and France (+3.4 per cent) than in Italy (+28 per cent), Germany (+2 per cent) or the

Production indices, 1990 = 100 (seasonally adjusted)

	1995	1996	1997	1997 fourth quarter	1997/96 % change
Industrial production					
Germany	98.0	97.7	99.7	100.5	2.0
Spain	103.2	102.5	109.5	112.8	6.8
France	100.1	99.5	102.9	106.5	3.4
Italy	107.8	104.8	107.7	109.9	2.8
United Kingdom	106.7	108.0	109.5	109.3	1.4
EU (15)	103.8	103.8	107.0	108.9	3.1
Metal-using industries					
EU (15)	103.4	102.0	110.2	113.3	8.0
of which:					
Motor vehicles	99.6	102.0	110.2	111.3	8.0
Mechanical engineering	97.0	96.8	99.1	101.0	2.3

Source: EUROSTAT, *Panorama of European Industry*; OECD, *Indicators of Industrial Activity*.

United Kingdom (+1.4 per cent). While the automobile industry was one of the principal beneficiaries of the upturn in the economy with an 8 per cent increase in activity, the construction industry only showed signs of recovery in a few Member countries. The most encouraging trends in this sector were to be found in the house-construction industry, which benefited from very low interest rates and the improved outlook for employment.

The economic growth observed in 1997 should probably continue and may well accelerate very slightly, with recent estimates predicting 2.7 per cent growth in GDP in the EU(15) area in 1998. Renewed optimism among industrialists and a sharp improvement in consumer confidence would seem to indicate that the general outlook for industrial activity in 1998 is very good. Despite the adverse impact of the financial crisis in South-East Asia on external demand, domestic demand should help to fuel healthy growth in activity in the main steel-consuming sectors. Activity in the construction sector is expected to pick up in response to the anticipated rise in investment. Output in the automobile sector is also expected to rise slightly compared with 1997. The electrical and non-electrical sectors should continue their upward trend, as in the past few years.

The economy should continue to grow in 1999 and GDP should rise by 2.8 per cent. Growth in investment should accelerate and might well increase by 4.9 per cent. Private consumption should also continue its upward trend, with anticipated growth of 2.6 per cent. The average rate of unemployment in the EU(15) area should continue to decline, falling to no more than 10.5 per cent of the working population. The upward trend in the level of activity of the main steel-consuming sectors should remain unchanged, as in 1998, although the increase may be slightly higher in the construction sector.

In the steel sector, apparent steel consumption in the EU(15) area (in finished product equivalents), after a sharp decline in 1996, rose by 12.8 per cent in 1997, an increase of 14.7 million tonnes. The real consumption of steel increased from 2.5 to 3.5 per cent due to the phenomenon of stock reconstruction which mainly happened during the two first quarters of 1997, within producers as well as traders and consumers.

Crude steel production increased by 13 million tonnes in 1997 in the whole EU, *i.e.* an increase of 8.8 per cent compared to 1996, to reach 159.9 million tonnes. Output rose in all 15 EU Member States, although the rate of increase varied from one country to another. The strong growth in demand led to a decline of just over 15 per cent in net exports from the EU. Imports rose by 6.6. per cent, while exports fell by 6.8 per cent over the same period. From the second quarter of 1997 onwards, prices for steel products, particularly flat products, began to pick up, rising by around 13 per cent over the year as a whole.

Demand for steel within the EU area should continue on an upward trend in 1998, driven by buoyant domestic demand, and consequently apparent steel consumption should rise by around 4.8 per cent. Crude steel production is expected to rise by 2.1 per cent to a new record level of about 163.4 million tonnes. The utilisation rate of production capacity should increase to around 83 per cent.

Yearly percentage changes in real and apparent steel consumption in the EU area

	1997/96 realised		1998/97 estimates		1999/98 forecast	
	Real	Apparent	Real	Apparent	Real	Apparent
Germany	−6.2	−4.8	4.0	4.0	−6.7	−6.1
Spain	11.2	22.7	8.5	4.4	10.7	10.6
France	1.4	11.5	10.0	7.7	7.3	5.4
Italy	19.6	33.0	11.7	6.8	4.5	4.9
United Kingdom	3.8	7.8	0.7	−1.4	2.9	3.6
Rest EU (15)	−5.1	6.0	12.0	1.4	2.8	4.2
Total EU (15)	2.3	10.2	8.1	3.9	2.3	2.7

Source: OECD Secretariat.

In view of these trends, net exports from the EU should fall by a further 18.5 per cent, accompanied by a 7.7 per cent increase in imports and a 6.6 per cent decline in exports. The sharp drop in imports to Asian countries, together with the glut of steel products – particularly from the NIS – in the market, give cause for concern over the trend in steel product prices in the second half of the year, given that the general trend during the first quarter had been upwards and would normally have been expected to level off in the second quarter.

Apparent steel consumption should continue to rise in 1999 by a further 3 per cent, slightly over 4 million tonnes more than in 1998. This continued rise in demand from the domestic market should lead to a further decline in exports of around 7.8 per cent, while imports should grow by no more than 1.4 per cent. Crude steel production looks set to increase by 1.1 per cent, which would take it past the 165 million tonnes mark, with a capacity utilisation rate of almost 85 per cent.

OTHER EUROPEAN COUNTRIES

This area comprises the Czech Republic, Hungary, Norway, Poland, Switzerland, Turkey and the former Yugoslavia. After the sharp fall in consumption reported in 1996, this area too experienced renewed growth in demand in 1997 and apparent steel consumption rose by 14.7 per cent, i.e. 3.5 million tonnes more than in 1996. Crude steel production rose by 8.8 per cent, slightly less than 3 million tonnes, and as a result net steel exports fell by 13.2 per cent due to a large increase in imports of around 15.2 per cent; at the same time, however, exports rose by 5.8 per cent.

In 1998, apparent steel consumption is expected to decline slightly, due to lower imports, while crude steel production should continue to rise by a further 0.5 per cent. Net exports of steel, on the other hand, should be up by 13 per cent.

1999 should see growth in steel consumption pick up pace throughout the area, with further growth of over 8.5 per cent. This increase will probably be accompanied by a rise in crude steel production of around 4.6 per cent and a 13.5 per cent reduction in net exports from the countries in this area, which would amount to 3.5 decline in imports and a 6.5 per cent decline in exports.

In **Hungary**, 1997 was an important year for the economy. While there had been encouraging signs by as early as 1996, it was not until 1997 that Hungary's economy took a sharp turn for the better with an increase in GDP of around 3.8 per cent and improvement in all of the main economic indicators. The economy was stimulated by soaring investment, which rose by 11 per cent, lower inflation and interest rates, the continued strength of net exports, a recovery in household demand after two years of decline and the first signs of a decline in unemployment. The basis for further economic growth in 1998 is now in place and growth in GDP is expected to accelerate to 4.3 per cent. Industrial output and investment should also start to pick up, leading to strong growth in the construction and mechanical engineering sectors. Inflation should fall substantially and could decline to around 14 per cent. Only unemployment will probably be slower to come down. The upward trend in economic activity should accelerate in 1999, with growth in GDP of the order of 4.6 per cent. Inflation should fall below the 10 per cent mark and investment should continue to rise by a further 11 per cent.

Apparent steel consumption grew strongly in 1997, rising by 14.6 per cent compared with 1996. Despite booming domestic demand and a small rise in steel exports, crude steel production fell by 9.6 per cent. This was because the Diosgyor mill changed its smelting technology to one based on scrap and had problems in procuring raw materials. In addition, output from the mill was affected, at least for a short period of time, by the change of ownership. The Ozd steel-making plant experienced similar problems, namely a change of owner and problems procuring raw materials, which resulted in lower output. The fall in steel production therefore led to a sharp increase, of around 58.6 per cent, in steel imports in order to meet demand.

Steel demand is expected to rise by a further 10 per cent in 1998. This increase will probably be accompanied by a 9.5 per cent rise in crude steel production, all of which should now be produced by continuous casting, and by a slight decline in steel imports. The acceleration in economic growth in 1999 should help steel consumption rise by at least 16 per cent, accompanied by a further increase of slightly over 8 per cent in crude steel production and of around 10 per cent in imports; steel exports should remain close to the 1997 and 1998 levels.

The **Norwegian economy** continued to grow vigorously in 1997, which it has every year since 1993, despite a slight slowdown early in the year. Growth in GDP over the year as a whole amounted to 4.0 per cent, while investment rose by over 10 per cent. A recovery in investment in the housing sector was accompanied by buoyant household consumption and, as a result, the economy was increasingly driven by domestic demand. The construction, shipbuilding and off-shore oil sectors all grew strongly. According to the most recent forecasts, overall economic growth should accelerate in 1998 and could possibly rise to 4.7 per cent, after which it should gradually fall back to 3.6 per cent in 1999 as growth in oil exports starts to level off. Traditional export products should nonetheless be stimulated by the economic recovery in Europe, and private consumption should remain strong.

Crude steel production rose by 11 per cent in 1997 and production plants operated at almost 100 per cent of their capacity. Apparent steel consumption rose by 7.7 per cent and net steel imports by 4.8 per cent. A slowdown in activity in the construction, shipbuilding and offshore oil sectors in 1998 should cause steel consumption to fall by around 5 per cent. Crude steel production may continue to rise only slightly, given that any increase will be limited by the number of plants working at 100 per cent capacity. Since steel exports are expected to rise by around 9 per cent, chiefly to the EU(15), steel imports should fall by about 3 per cent. In 1999, steel consumption is expected to pick up by around 3.8 per cent, with production plants still working at maximum capacity, and steel exports should fall by about 10 per cent.

Poland's economy continued to grow at a high rate for the sixth year running. GDP grew by 6.9 per cent in 1997 and disinflation proceeded at a steady and regular pace. The acceleration in economic growth in 1997 is attributable to faster growth in valued added in the industrial sector (9.7 per cent in 1997) and very strong growth in activity in the construction sector, which expanded by 17.6 per cent in 1997. The growth in the economy had a highly positive impact on employment and the rate of unemployment fell from 13.6 per cent in 1996 to 10.5 per cent in 1997. Private consumption continued to grow rapidly, rising by 7 per cent in 1997, and investment rose by a further 21.9 per cent. Total industrial output grew by 11.2 per cent, and activity in the investment and consumer goods sectors increased by 16 per cent and 13 per cent respectively. All the main steel-consuming sectors reported very large increases in activity, ranging from around 18 per cent in the automobile sector, to 17.2 per cent in the machinery and machine-tool sector and 23-35 per cent for other transport equipment.

The continuing high levels of investment may well be tempered by the increase in real interest rates, but have provided the basis for a rapid and sustained increase in production in 1998 and 1999. Investment is therefore estimated to grow by a further 15.1 per cent in 1998 and 12.5 per cent in 1999. GDP should grow at a rate of 5.8 per cent in 1998 and 5.6 per cent in 1999. Private consumption will probably remain buoyant but should gradually taper off. The rate of unemployment should continue to fall, although at a moderate pace, and inflation should be brought down below 10 per cent in 1999.

The excellent performance of the Polish economy in 1997 and strong growth in output from the main steel-consuming sectors led to an increase in steel consumption of around 18.7 per cent, slightly over 1 million tonnes more than in 1996. Crude steel production rose by 11.1 per cent to a level of 11.6 million tonnes. Since steel exports, too, grew by 10.2 per cent, steel imports rose by around almost 23 per cent. 1998 should see steel demand continue to rise, although at a slightly lower rate of 4.8 per cent. Growth in crude steel production, on the other hand, will probably be limited to 1 per cent, and steel imports, up by 17.2 per cent, should allow demand to be met. In 1999, the positive trend should continue with steel demand increasing again by over 4 per cent, while steel production should rise by 3.8 per cent.

In **Switzerland**, after declining in 1996, the economy started to grow again in 1997 and GDP rose by 0.7 per cent. A strong export performance, aided by a fall of around 7 per cent in the value of the Swiss franc, coupled with rising household consumption helped to boost domestic demand. The continuing crisis in the construction sector, however, led to a further decline in the gross fixed capital formation of around 1.5 per cent. In 1998, growth in the economy is expected to accelerate and GDP is forecast to rise by 1.5 per cent. Despite stagnation in the construction sector, the business climate is expected to be better than in 1997 and, due to the favourable economic climate, growth is expected in the supply of speciality steels to the automobile sector and in the machinery and instruments sector. In 1999, the rate of growth should gradually accelerate to 1.8 per cent, accompanied by growth in household consumption, while investment should start to recover, rising by 5 per cent.

Matching the trend in the economy, steel demand rose by 18.9 per cent in 1997 to a level of 2.1 million tonnes of finished product equivalents. Crude steel production rose by 22.6 per cent, compared with 1996, to a total of 1.1 million tonnes. This increase was partly attributable to the partial shutdowns in production in 1996 to allow maintenance and upgrading work to be carried out. Production units operated at almost 100 per cent capacity in 1997. With regard to foreign trade, steel imports increased by 18 per cent, compared with 1996, to 1.9 million tonnes. Exports too rose by 24 per cent to 0.9 million tonnes. As a result, net steel imports rose by around 12.5 per cent.

In 1998, steel demand is forecast to grow by around 3 per cent. This increase should be accompanied by a relatively small increase in crude steel production, limited by the number of units operating at full capacity. Against this background, a slight improvement in prices should allow Swiss steel-makers to achieve a satisfactory profit level. The same rising trend in steel demand should continue in 1999, although it may be accompanied by a fall in exports of about 10 per cent due to the constraints on the supply side.

Growth in the economy of the **Czech Republic** has been gradually slowing since 1995 and in 1997 GDP grew by no more than about 1 per cent. It is worth bearing in mind, however, that the Czech economy was badly affected by the catastrophic floods in 1997. Exports were boosted by the 10 per cent devaluation of the Czech crown, while imports grew rapidly in response to increased household demand. The rate of inflation slowed to around 8.5 per cent in 1997, although unemployment rose from 3.5 per cent in 1996 to 5.2 per cent in 1997. Investment was down by 4.9 per cent. The economic slowdown looks set to continue in 1998 and 1999. GDP will probably grow by no more than 0.9 per cent in 1998, rising to 1.2 per cent in 1999. Further to the stabilisation measures recently introduced, private consumption will probably decline while high interest rates should continue to reduce private investment in 1998. The situation should nonetheless start to improve from 1999 onwards.

Despite the slowdown in economic growth, steel demand rose by 13.5 per cent in 1997 but still remained 6 per cent below the level reached in 1995. Crude steel production increased by 3.7 per cent to 6.8 million tonnes, and trade in steel also experienced a similar rising trend; net exports, however, fell by around 4 per cent. In 1998 and 1999, the moderate trend increase in steel consumption should continue; crude steel production should fall off slightly in 1998, due to a fall in exports, but may pick up slightly in 1999, particularly if imports start to fall.

Turkey's economy continued to grow strongly, expanding by 8.0 per cent in 1997, although macroeconomic imbalances continue to give cause for concern. Growth was stimulated by increased public expenditure, but the rate of growth in overall domestic demand and investment slowed to the region of 6 per cent, as did private consumption. Inflation reached 82 per cent. Apparent steel consumption rose by about 12.6 per cent and crude steel production by 6.7 per cent. Imports increased by 11.3 per cent to meet the increase in demand, while exports were down by 3.7 per cent.

Depending on the budget decisions taken by the government, GDP growth may slow to no more than 5.5 per cent in 1998, and 5 per cent in 1999, as the impacts of a tightening of monetary policy and budget austerity begin to make themselves felt on expenditure in the private sector. Apparent steel consumption could maintain the previous year level in 1998, accompanied by a 2.3 per cent contraction in crude steel production, a 8.3 per cent decline in steel imports and a fall of around 12.5 per cent in steel exports, given that the traditional market for a quarter of all Turkish steel exports are the countries of South-East Asia. 1999 may see a small increase of just over 1 per cent in steel consumption, and production could well rise by 3.9 per cent.

JAPAN

Growth in Japan's economy slowed down in 1997 and GDP grew by merely 0.9 per cent, this slowdown being primarily attributable to cuts in spending on public works and a reduction of 3.4 per cent in total fixed investment, although private investment in the manufacturing sector remained stable. Private consumption slowed, rising by merely 1.1 per cent, due in part to uncertainty in the household sector over the outlook for the economy. Lastly, with final domestic demand slightly down, stock inventories were high and production started to fall.

Despite the downwards trend in the house-building sector, industrial output rose by 4 per cent in 1997. Although private car sales were stagnant in Japan, output in the automobile sector, driven by demand for exports, increased by 8.1 per cent. Production in the machinery and machine-tool sector and the electrical equipment sector increase, private industrial investment remained steady and export demand grew particularly strongly. Activity in the shipbuilding sector, in contrast, stagnated.

In order to reverse the gloomy economic outlook for households and industry and to return the economy to a stable cycle of growth driven by private demand, the government has decided to introduce a 16 000 billion yen recovery plan, although it is not as yet considering any reductions in various charges and taxes. Despite these efforts, the most recent economic forecasts for 1998 indicate a fall in GDP of around 0.3 per cent over the year. The main indicators are expected to show a downward trend, private consumption should decline by 0.4 per cent, investment by 2.3 per cent and final domestic demand by 1 per cent. The same forecasts also predict that the economy will start to recover in 1999 with 1.3 per cent growth in GDP and a similar increase in private consumption.

In the steel sector, apparent consumption rose by 2.9 per cent in 1997, an increase of 2.2 million tonnes over the previous year. Due to some rebuilding of stock inventories, steel consumption in real terms only increased by around 1.6 per cent. Owing to the problems that the Japanese economy will have to face in 1998, apparent steel consumption is likely to experience a further decline of around 4.6 per cent before rebounding in 1999 with estimated growth of about 3 per cent.

Increased demand for steel and stock rebuilding led to a 5.8 per cent increase in crude steel production in 1997, 5.8 million tonnes more than in 1996. The capacity utilisation rate was 70 per cent. In 1998, a decline in domestic demand, and probably exports too, should result in a significant reduction in crude steel production of the order of 7.2 per cent, with the volume of production falling by 7.6 million tonnes to around 97 million tonnes. In 1999, depending upon whether demand recovers as expected, crude steel production may begin to rise again and with growth of around 4 per cent could well break the 100 million tonnes mark.

In 1997, despite the lower yen, steel imports rose by 8 per cent and the share of imports in the Japanese market increased from 7.5 per cent in 1996 to 7.9 per cent in 1997. At the same time, exports of Japanese steel rose by 18.6 per cent to 22.8 million tonnes. Although there may have been fears that the economic and financial crisis that gripped South-East Asia from the summer onwards would have a major impact on Japanese steel exports, the latter appear to have only been marginally affected in 1997. Consequently, although steel exports to Thailand were slightly down on the previous year's levels,

Indices of activity in the steel-consuming sectors, 1990 = 100

	1995	1996	1997	1997/96 % change
Industrial production	96	99	103	+4.0
Production passenger cars	85	86	93	+8.1
Production commercial vehicles	87	89	95	+6.7
Non-electrical machinery	81	86	91	+5.8
Electrical machinery	116	125	137	+9.6
Shipbuilding	113	121	121	0.0

Source: OECD, Indicators of Industrial Activity.

those to the ASEAN area as a whole were up by 13.5 per cent. Exports to Korea reached a record level of 3.6 million tonnes, an increase of 6.6 per cent on 1996. Exports to China increased by 4.8 per cent, to Chinese Taipei by 8.8 per cent and to the United States by 31.9 per cent.

In 1998, on the basis of estimated demand for steel, imports are expected to contract by around 20 per cent. Although exports continued to rise during the first two months of the year, particularly those to the United States, exports in 1998 are expected to fall relatively significantly, particularly those to the Asian region. In contrast, exports in 1999 may well start to rise again, while imports should continue to decline slightly.

KOREA

In November 1997, the Korean economy suffered a severe monetary crisis which led to the implementation of an aid programme by the IMF. In return for the agreement with the IMF, the Korean government is drawing up a vast programme of reform aimed at restructuring the financial and business sectors and at creating a more flexible labour market. In addition, the government is due to move faster on opening up the Korean market to foreign investors and to implement restrictive macroeconomic policies by reducing the money supply and public expenditure.

In 1997, GDP growth slowed to no more than 5.5 per cent. The onset of the crisis led to a fall in investment of around 3.5 per cent and a slowdown in private consumption. Unemployment began to rise and finished at 2.6 per cent over the year. Exports of goods and services soared by 23.6 per cent, while imports grew by merely 3.8 per cent. The economic situation in Korea will continue to be extremely difficult in 1998. Provided that the vast programme of reforms introduced by the Korean government is rapidly implemented, the conditions should be right for the economy to become competitive again. The most recent forecasts indicate that GDP will contract by 0.2 per cent and that private consumption will decline by 4 per cent, public expenditure by 5 per cent and total domestic demand by 18.7 per cent. Investment is expected to fall by 18.7 per cent. Exports should rise by a further 12 per cent, but imports are likely to fall by 6 per cent. Unemployment and inflation should start to rise significantly. However, there should be a marked recovery in 1999 when GDP should grow by 4 per cent. Private consumption should start to pick up, while public expenditure and investment should decline a little further.

Activity in the construction sector, which rose by 2.5 per cent in 1997, should fall by 10 per cent in 1998, the first decline since 1980. After a 3.7 per cent reduction in output from the automobile sector, 1998 is expected to see a further decline of 5.3 per cent as a result of the sharp fall in car sales in the domestic market. The electronic consumer goods industry saw a 2.3 per cent fall in sales in 1997 and should expect a further 1.5 per cent fall in 1998, since domestic demand is expected to contract by 5.6 per cent as a result of lower private consumption. In the shipbuilding sector, new orders in 1997 were 83.5 per cent up on the previous year. In 1998, in view of the large number of orders already placed, activity in this sector should remain at more or less the same level as in 1997.

Apparent steel consumption in 1997 increased by merely 1 per cent to 37.7 million tonnes. This low rate of growth was the result of the sharp increase in stocks in 1996. Crude steel production rose by 9.4 per cent, i.e. 3.7 million tonnes more than in 1996, to a new record level of 42.5 million tonnes. The capacity utilisation rate was 93 per cent. As a result of these trends, steel imports fell by 15.9 per cent while exports rose by 12.6 per cent.

Because of the economic crisis, apparent steel consumption in 1998 is likely to fall by 13.2 per cent to no more than 32.7 million tonnes, 5 million tonnes less than in 1997. The slowdown in activity in the main steel-consuming sectors and a sharp decline in imports, including semi-finished products, as well as the difficulty in finding new outlets for exports should lead to a decline in crude steel production of around 5.5 per cent. The 3.4 mtpa increase in production capacity, coupled with falling production levels, should lower the utilisation rate to less than 82 per cent. Steel imports are likely to fall by 18.4 per cent, whereas exports, despite fierce competition, should rise by around 10 per cent.

It is possible that steel consumption may stage a limited recovery in 1999, growing by around 3.3 per cent. This recovery will probably be accompanied by a fall in net steel exports of about

6.3 per cent, although crude steel production should rise by around 2 per cent. If the planned new production units due to come on stream in 1999 are completed, the capacity utilisation rate could well fall to around 76 per cent.

AUSTRALIA AND NEW ZEALAND

The **Australian** economy continued to grow in 1997, but at a lower rate than it had over the previous six years. GDP grew by 2.7 per cent, and inflation remained very low at around 1.5 per cent. The economy was buoyed by investment, which was 10 per cent up, and by private consumption which increased by 3.5 per cent. Industrial output was merely 1.1 per cent up, but activity in the construction sector remained firmly on course, thanks in particular to more than affordable house prices. In 1998 and 1999, the economy should start growing slightly faster and GDP should increase by 3.2 per cent a year. Investment should remain strong, rising by 7.7 per cent in 1998 and 4.5 per cent in 1999. Private consumption will probably rise by over 4 per cent in 1998 and unemployment should gradually decline over the period to no more than 7.7 per cent in 1999.

1997 was a record-breaking year on all fronts for the Australian steel industry. Steel production, productivity, apparent consumption and exports all reached record levels. On the other hand, however, imports too reached record levels. Apparent steel consumption amounted to 5.8 million tonnes, up 8.7 per cent on the 1996 level. Crude steel production, at 8.7 million tonnes, was 3.7 per cent higher than the previous year's total. The Asian crisis of 1997 made relatively little impact on either production, consumption or exports of Australian steel. The impact they did have, however, was in the form of increased imports from Asian countries.

Steel imports rose 24 per cent to 1.2 million tonnes and started to accelerate during the second half of the year. Imports were up by 183 per cent from ASEAN member countries, by 137 per cent from Chinese Taipei, by 47 per cent from China and by 40 per cent from Korea. Imports from South Africa, in contrast, fell by 40 per cent. The average price per tonne of imported steel fell by 7.5 per cent. Steel exports rose by merely 1.9 per cent, but to a new record level of 3.3 million tonnes. Exports to ASEAN member countries were up by 39 per cent, to Korea by 31 per cent and to Hong Kong by 46 per cent. In contrast, exports, to Chinese Taipei, China, Japan, Canada and the United States declined. Exports to Europe, the Middle East and Latin America rose, in some cases quite substantially, but only accounted for a small volume of sales. Exports of coated flat products continue to be affected by anti-dumping legislation in the United States and Canada.

In 1998, apparent steel consumption looks set to rise by a further 3.4 per cent to 6 million tonnes. Exports are likely to fall by 2.4 per cent and crude steel production should increase by around 2 per cent. The capacity utilisation rate is likely to be 97 per cent. In 1999, crude steel production will probably decline by 2.2 per cent, but because of cuts in capacity the utilisation rate should remain at 97 per cent. Despite an increase of around 16 per cent in imports and of 3 per cent in exports, apparent steel consumption is likely to be 0.8 per cent lower.

New Zealand's economy should follow a broadly similar trend to that of the Australian economy. With regard to the steel sector, crude steel production fell by 6.2 per cent in 1997, which, combined with a 12.5 per cent increase in exports, led to a decline in apparent steel consumption of 14.3 per cent. In 1998 and 1999, steel production is expected to rise by 5-6 per cent a year and steel consumption at the slightly higher rate of 7-8 per cent.

BRAZIL

Brazil's economy grew by 3.0 per cent in 1997 and inflation remained astonishingly low, falling to below 5.2 per cent. However, since October 1997 the Brazilian currency has been under heavy pressure resulting from the flight of short-term capital as the effects of the financial crisis in South-East Asia started to make themselves felt more widely. To counter this problem, interest rates have been raised substantially. Fiscal reforms have been introduced with the aim of reducing public deficits. In 1998, these measures should slow the economic growth rate to no more than 2 per cent, although it should rise back to 4 per cent by as early as 1999. Revenue from privatisations, which are expected to raise 55 billion dollars over the period 1998/99, should help bring about a significant reduction in public debt. Inflation will probably continue to fall and should amount to no more than 4 per cent, while private consumption should begin to rise.

Apparent steel consumption would seem to have grown by over 18.6 per cent in 1997 to a new record level of 15.4 million tonnes of finished product equivalents. Net exports of steel products fell by 15.2 per cent in response to strong growth in imports and a drop of around 10 per cent in exports. Crude steel production rose by merely 3.6 per cent, which nonetheless amounted to 1 million tonnes more than in 1996.

In 1998, the steep rise in interest rates and the slowdown in economic activity should lead to a mere 3 per cent increase in apparent steel consumption. Steel imports should fall by 62 per cent to around the level reported in 1995. Steel exports may well rise by 4.8 per cent, due in large part to increased exports to North America. This should help fuel a further 1 per cent increase in crude steel production. The improved economic climate in 1999 should allow apparent steel consumption to rise by around 6.3 per cent. Crude steel production could well rise by 2.4 per cent, thus passing the 27 million tonne mark. Although steel exports will probably remain at the 1998 level, imports may have to double to meet demand.

OTHER NON-OECD ECONOMIES

Other Latin American countries

The economy of the region as a whole grew strongly in 1997, with particularly strong growth in Argentina, Chile and Peru and a return to growth in Venezuela after the recession of 1996. Inflation, which was still over 10 per cent in Colombia and Venezuela, nonetheless fell in almost all the countries in the region. By contrast, inflation in Argentina was very slightly up on its previously very low level as a result of vigorous growth in the economy. Investment in the central and southern countries of the region responded favourably to the new possibilities opened up by the liberalisation of trade, falling interest rates and increased consumer demand. In particular, direct foreign investment appears to have been in excess of 25 billion dollars. The economy of the region as a whole was apparently over 5 per cent in 1997 and should continue at the same pace in 1998 and 1999.

The economic trend created a highly favourable climate for the steel industry in Latin America and apparent steel consumption rose by 7.7 per cent in 1997. Net imports of steel to the area as a whole was sharply up, an increase of 383 per cent compared with 1996, but this was primarily due to a steep decline in exports since in reality imports only rose slightly. Crude steel production rose by 5.7 per cent to 12 million tonnes and demand for steel in 1998 may well start to pick up and could rise by 20 per cent to 13.5 million tonnes. As a result of this increase, crude steel production, which will probably rise by merely 0.6 per cent, will not be able to meet demand, which in turn should lead to a sharp increase of 41.5 per cent in steel imports and a 15 per cent decline in exports. The rising trend in steel demand should continue in 1999, when apparent consumption should rise by 9.5 per cent. An increase in steel production to 13 million tonnes, an 8.5 per cent increase compared with 1998, should keep imports down to merely 3 per cent, whereas exports should experience a further decline of around 6.5 per cent.

Africa and the Middle East

In **South Africa**, apparent steel consumption, which had grown by 6.3 per cent in 1997, should continue on an upward trend over the next two years, rising by 1.5 per cent in 1998 and by around 3.5 per cent in 1999. In 1997, crude steel production rose by 3.9 per cent to 8.3 million tonnes. It is thought that it will probably rise by a further 2.3 per cent in 1998 and by 5.5 per cent in 1999. Imports to, and exports from, South Africa may fall in 1998 but should start to rise again in 1999.

Demand for steel in the rest of the **African continent** fell by 3.4 per cent in 1997 and would appear to have fallen once again below the 4 million tonnes mark. Crude steel production dropped sharply for the sixth year running and the 20.3 per cent fall reported in 1997 brought production down to less than 1 million tonnes. Steel imports account for almost 90 per cent of the steel consumed in Africa, and exports have always been extremely low and primarily limited to trade within Africa. In 1998 and 1999, a modest recovery is expected in steel demand, which will nonetheless remain at a low level. Steel production should continue to fall in 1998 and demand will be met through increased imports, although production may start to rise again in 1999.

In the **Middle East**, steel consumption, which had risen for ten years in succession, continued to grow significantly and in 1997 rose by 8.5 per cent to 28.7 million tonnes, 2.3 million tonnes more than in 1996. Crude steel production also continued to rise, and the 7 per cent increase reported in 1997 brought output up to 13.5 million tonnes. With the commissioning of 2.6 mtpa of additional capacity, the average utilisation rate of capacity was 72 per cent. Steel imports rose by 9.2 per cent to 17.5 million tonnes. Exports, too, rose by 5.3 per cent but in volume terms amounted to no more than 1 million tonnes. In 1998, apparent steel consumption in this region should slow to no more than 0.5 per cent, before growing at a stronger rate of around 1.2 per cent in 1999. Crude steel production is expected to fall by 6.9 per cent in 1998 and may well contract by a further 0.8 per cent in 1999. Steel imports should rise by over 1 million tonnes in 1998 and by a further 1.3 per cent in 1999.

Asia

The Asian continent, excluding China, can be divided in three separate regions: India, the five main members of the ASEAN, defined as ASEAN(5) and namely Indonesia, Malaysia, the Philippines, Singapore and Thailand, and lastly all other Asian countries, including Chinese Taipei, Vietnam and North Korea, which in the past was covered at the same time as China.

In **India**, after three years of rapid economic growth at a rate of over 7 per cent, the economy started to slow somewhat in 1997 and GDP grew by merely 5 per cent. Industrial output continued to rise and was 4.9 per cent up, significantly below its potential. Exports and imports also grew more slowly than expected. The rupee, because the Indian economy is less open than neighbouring economies, fared less badly than other currencies in the region, but despite a rise in interest rates of two percentage points lost slightly more than 10 per cent of its value against the dollar between August and December 1997. The rise in interest rates, on the other hand, had an adverse impact on investment, which was already at a low level.

The less favourable economic climate in 1997 led to a slowdown in apparent steel consumption, which grew by merely 0.4 per cent. At the same time, crude steel production remained at the same level as in 1996. Net steel imports rose slightly, with imports growing slightly more than exports. Demand for steel in 1998 should fall by slightly over 5 per cent, leading to a decline in net steel imports, and crude steel production by 5.5 per cent. In 1999, steel consumption will probably remain stable if not slightly up on the 1998 level. Crude steel production could well increase by around 4.5 per cent and steel exports could rise to the point where, combined with a fall in imports, India could become, although only by a whisker, a net exporter of steel.

Of the five **ASEAN(5)** members that appear to have been the worst affected by the financial crisis which began during the summer of 1997, Thailand was the first economy to be affected and the country where the slowdown in economic growth was most marked. However, the impacts of this crisis were rapidly felt by the other economies in the region, of which the worst affected were Indonesia, Malaysia, the Philippines, Singapore and, albeit to a lesser extent, Chinese Taipei. While it is not possible to give a detailed account of all the impacts of the crisis on the economies of these countries, it is nonetheless worth noting that in 1998, although there should at best be zero growth in the economies of Indonesia and Thailand, there could well be growth of the order of 3 per cent in Malaysia and the Philippines. While measures have been taken to promote foreign investment, and privatisation should accelerate, there is every indication that domestic demand will be sharply down and that budgetary restrictions will affect infrastructure projects and business activities. Nonetheless, there is hope that the measures that governments will have to introduce, provided that they are actually implemented, will allow economic growth to resume, no doubt gradually, by as early as 1999.

With regard to the steel sector, it is worth recalling first of all that steel consumption has been growing strongly in these countries since the mid-1980s, rising from 17.3 million tonnes in 1990 to 34.2 million tonnes in 1996. Crude steel production has followed a similar trend, rising from 5.8 million tonnes in 1990 to 10.9 million tonnes in 1996. The growth in output, however, has not been sufficient to keep pace with strong demand and has led to a sharp increase in imports, which more than doubled between 1990 and 1996 to reach 27.8 million tonnes in 1996. It should also be noted that a large share of these imports consisted of semi-finished products, which accounted for close on 38 per cent of all steel imports. The growing importance of these five countries in the world steel market can be seen in the fact that, with total steel imports of 27.8 million tonnes in 1996, all five countries taken together can be considered to be the world's largest importer of steel, outranking even the United States which in 1996 imported only 26.9 million tonnes of steel. The rapid growth in steel consumption has stimulated investment in the steel industry in this region, particularly since the early 1990s. Most planned investment projects will be operational by the turn of the century and the increase in the steel production capacity of the ASEAN(5) members between 1995 and 1999 should be excess of 11 mtpa.

The economic crisis in 1997 led to an 8.1 per cent fall in steel consumption in these five economies as a whole. However, while steel demand fell in all five countries, the size of the decline varied from one country to another, from 2.9 per cent in Indonesia to 8.7 per cent in Malaysia, 13.4 per cent in the Philippines, 2.3 per cent in Singapore and 10.6 per cent in Thailand. Crude steel production in the ASEAN(5) area also declined but to a lesser extent, falling by 2.7 per cent. The largest drop was in imports which fell by 13 per cent to 24.2 million tonnes, 3.6 million tonnes less than in 1996. At the same time, steel exports from these countries, which were primarily to other Asian countries, plummeted by 32.3 per cent.

In 1998, the year during which the effects of the crisis should make themselves felt more keenly, apparent steel consumption is expected to fall by 26.5 per cent. The size of the decline will vary from one country to another, but will be substantial in all of them. Crude steel production in the area should also fall by 16 per cent, except in Singapore whose output, less than 0.4 million tonnes, is already very low. Steel exports are expected to decline by a further 11.5 per cent, whereas imports should be slashed by a further 30 per cent, 7 million tonnes less than in 1997.

In 1999, the trends in the steel market in these five countries will clearly depend upon the effectiveness of the measures taken to remedy the crisis and upon whether or not there is a return to growth in the economy and investment. In the most optimistic scenario, in which the crisis is overcome relatively quickly, steel consumption could rise by around 5 per cent. Not all the economies of the countries in this group will recover, however. Even in the best-case scenario, consumption is expected to continue to decline in Thailand, would at best remain stagnant in Indonesia, but could well start to increase in Malaysia, the Philippines and Singapore. This recovery is likely to be accompanied by a proportionately similar increase in steel production, but probably not in Indonesia where production is expected to continue to decline. There might also be a modest pick-up in steel imports of the order of just over 1 million tonnes.

Trends in the economies of the **other Asian countries**, of which the most important with regard to steel are Chinese Taipei, Pakistan, North Korea and Vietnam, varied substantially in 1997. North Korea has been facing economic difficulties for several years now. Chinese Taipei was affected by the Asian crisis, but to a lesser extent than the countries reviewed above, and its economy, despite the fall in the value of its currency, would appear to have grown in 1997, buoyed by strong growth in private consumption.

Steel consumption in these countries as a whole grew by a further 12.9 per cent in 1997 to 31.2 million tonnes. It should be noted that Chinese Taipei alone accounted for 21.2 million tonnes, *i.e.* 68 per cent of total consumption in the area. Steel consumption in Chinese Taipei rose by 15.6 per cent in 1997. Crude steel production in the area rose by 29.5 per cent to over 17 million tonnes, but here again 94.7 per cent of this production, *i.e.* 16.2 million tonnes, were accounted for by Chinese Taipei alone. Net imports of steel to the area declined by 8.9 per cent. 1998 should see a fairly sharp decline in steel demand, primarily due to the downward trend in intra-regional trade, driven by currency devaluations in the ASEAN(5) countries. Steel demand should nonetheless remain at a satisfactory level in Chinese Taipei, where crude steel production should rise by a further 1.2 per cent. Moreover, production in the area as a whole should rise by a further 1.8 per cent. The overall volume of steel trade (both imports and exports) should be fairly significantly down. In 1999, steel demand should start to rise again by around 5.3 per cent. There will probably be no more than a modest increase in steel production and the increased demand will therefore have to be met by imports, which will probably rise at a slightly higher rate.

Central and Eastern Europe

Since 1996 when the Czech Republic, Hungary and Poland became Members of the OECD, the only countries left in this area are Albania, Bulgaria, Romania and the Slovak Republic. The steel market in Albania can still be considered to be flat. For the seventh year running crude steel production in Albania remained at 0.02 million tonnes and no significant change in this situation is foreseen over the next few years.

The **Bulgarian** economy, which had contracted by 10.9 per cent in 1996, declined by a further 8 per cent in 1997. The deep recession into which the country has been plunged resulted in a sharp devaluation of the Bulgarian currency against the dollar. Inflation accelerated significantly to reach 580 per cent for 1997, but following the agreement reached with the IMF on 1 July slowed dramatically to no more than 18 per cent during the latter half of the year and to merely 0.5 per cent during the last two months of 1997. As a result of the astonishingly rapid success of the stabilisation and structural reform programme, interest rates rapidly came down in the second half of the year. Unemployment increased, rising from 12.5 per cent in 1996 to 14 per cent in 1997. Privatisation of state-owned companies is still proceeding very slowly and it is estimated that in 1997 a mere 25 per cent of Bulgarian companies had been privatised – and not one of the three state-owned steel companies. 1998 is expected to see economic activity pick up and GDP could well grow by 4 per cent, a recovery that should start to gather pace in 1999.

Apparent steel consumption, which had fallen by 47.2 per cent in 1996, recovered in 1997 and grew by 70.7 per cent, although at 0.7 million tonnes nonetheless remained at a very low level. Crude steel production was up by 6.9 per cent, although stocks held by producers appear to have risen substantially in that only 84 per cent of the steel produced was sold. Steel exports rose to 1.5 million tonnes, 62 per cent of which was exported to the EU(15). Imports increased, but still remained at a very low level of 0.2 million tonnes. In 1998 and 1999, steel consumption is expected to rise by 14 per cent and 31 per cent respectively. The increase in steel production should be far more modest, between 2.2 per cent and 2.5 per cent a year, and as a result steel exports are expected to fall.

Growth in GDP in the **Slovak Republic** remained buoyant in 1997 at 5.8 per cent, although this was slightly down on the 6.9 per cent growth achieved in 1996. The economy should continue to grow at a rate of 5.5 per cent in 1998 and of 5 per cent in 1999. Domestic demand, boosted by strong growth in real wage levels and an expansionary budgetary policy, was the main engine for growth. In July, the government reintroduced a 7 per cent surcharge on imports which will be gradually reduced to 5 per cent from January to April 1998, then to 3 per cent from May to September and will be completely phased out on 1 October 1998. Inflation, which was running at 6.1 per cent in 1997, should remain at 6 per cent in 1998 and 1999. The rate of unemployment, which had risen to 13 per cent in 1997, should fall to 12 per cent in 1998 and then 11 per cent in 1999.

Apparent steel consumption rose by 15.5 per cent in 1997 and should increase by a further 1.8 per cent in 1998 and then 10.5 per cent in 1999. Crude steel production rose by merely 5.3 per cent in 1997 to a total of 3.8 million tonnes. It should rise by a further 1.3 per cent in 1998 and then 1.6 per cent in 1999. Steel imports fell by 4 per cent in 1997 but could well rise by 5.6 per cent in 1998 before experiencing a further decline of 7.9 per cent in 1999. Exports, which had fallen by 0.7 per cent in 1997, should be slightly up in 1998 but will probably start to decline again in 1999.

The measures introduced in **Romania** in 1996 to stabilise the economy and accelerate the process of structural reform led to a 6.6 per cent decline in GDP in 1997. Inflation soared to 151 per cent and the value of the currency fell by almost 50 per cent against the dollar. Industrial output was substantially down by around 20 per cent. Unemployment rose sharply to almost 9.5 per cent, due to the downturn in activity in the coal-mining, mechanical engineering, petrochemical and civil engineering sectors, and also due to the restructuring of state-owned enterprises. The sectors of the economy that nonetheless made progress included telecommunications, agricultural production and steel-making. 1998 is expected to see zero growth in GDP, 45 per cent inflation and unemployment rise to over 12 per cent. A small increase of 2-3 per cent in economic activity is expected in 1999.

The poor economic climate led to a 7.3 per cent decline in apparent steel consumption in 1997, and in volume terms consumption fell back below the 3 million tonnes mark. Steel consumption should increase by 5.2 per cent in 1998 and then accelerate to 10.5 per cent in 1999. Crude steel production was up by 9.7 per cent in 1997 and should rise by 2.8 per cent in 1998 and by a further 5.7 per cent in 1999. Steel imports, primarily semi-finished products from the NIS, which had increased by 12.7 per cent in 1997, should continue to rise at a similar pace in 1998 before levelling off in 1999. Steel exports, which had soared by 34.2 per cent in 1997, should rise by a further 2.6 per cent in 1998 and then remain at that level in 1999.

New Independent States

The long period of contraction in the Russian economy finally came to an end in 1997. Against a background of falling inflation and a sharp reduction in interest rates, GDP grew by 0.4 per cent. Industrial output rose by 1.9 per cent in industry as a whole and by 3.5 per cent in the manufacturing sector. Only investment continued to lag behind. In 1998 and 1999, the recovery should start to gather pace with growth in GDP of 3 per cent in 1998 and 5 per cent in 1999. Ukraine has again made progress in stabilising its economy. Inflation fell to 12 per cent, with unemployment still remaining at the very low level of 2 per cent. Despite this progress, GDP declined by a further 4 per cent in 1997. The pace of structural reform, and particularly privatisations, remained sluggish throughout the year.

In the NIS area as a whole, apparent steel consumption recovered in 1997, rising to 9.7 per cent compared with 1996. Apparent steel consumption in Russia rose by merely 1 per cent, but in Ukraine and the other NIS rose by 23.5 per cent and 21.7 per cent respectively. Crude steel production increased by 3.5 per cent to 79.8 million tonnes. The 1.5 per cent decline in production in Russia was offset by a 6 per cent increase in Ukraine and by a 17.7 per cent increase in the other NIS. Steel imports to the NIS area rose by 10.2 per cent in 1997 and exports by 1.8 per cent.

In 1998, total apparent steel consumption in the NIS should increase by 18.4 per cent and may well reach a total of 30 million tonnes, rising by 12.1 per cent in Russia, 27 per cent in Ukraine and 23.6 per cent in the remaining States. Imports may well fall by 1.3 per cent, but exports, due to lower demand in Asia, could fall sharply by 12 per cent, *i.e.* 5 million tonnes less than in 1997. As a result, crude

Capacities and crude steel production in 1997

	Capacity	Production	Utilisation rate %
	In thousand tonnes		
Belarus	1 200	1 110	92.5
Kazakhstan	4 000	3 825	95.6
Federation of Russia	91 600	48 400	52.9
Ukraine	60 800	24 740	40.6
Uzbekistan	2 000	365	18.3
Total CIS	159 600	76 395	47.9
Azerbaïjan	800	50	6.3
Moldova	1 000	811	81.1
Georgia	1 800	110	6.1
Total NIS	163 800	79 800	48.7

Source: OECD, *Two-yearly Report on Steel Production Capacity in Non-member Countries.*

steel production in the NIS may decline by 0.4 per cent, with Russia accounting for the entire decline. In 1999, steel demand should continue to rise by around 15 per cent in the NIS area as a whole, with crude steel production rising by 4 per cent and exports experiencing a further decline of around 5 per cent.

China

In China, growth remained buoyant in 1997 and GDP grew by 9.2 per cent, a lower rate of growth than in previous years. A restrictive monetary policy and tight restraints on lending, coupled with a good harvest, helped to keep inflation down to a low level of around 3 per cent. A strong export performance stimulated growth in 1997, but the domestic economy showed signs of slowing; high interest rates and weaker growth in real wages also helped to dampen domestic demand. Interest rates were reduced in October in order to stimulate growth in consumer demand and thus encourage investment. The Chinese economy will probably continue its soft landing in 1998 and 1999, in response to the primarily restrictive stance of China's economic policy. Inflation should remain at a relatively low level. More reasonable policies towards foreign investment might possibly be introduced, and the reform of the banking system and state-owned enterprises should start to gather pace. The major challenge that will have to be confronted at a social level over the next few years, however, will be to ensure the smooth transition to new jobs of the millions of workers who will lose their jobs as a result of bankruptcies, restructuring, mergers or privatisations in the public sector. It is possible that by as early as 1998 some 11 million workers may lose their jobs.

Following the strong recovery in steel consumption in 1996, a further increase of 3.9 per cent was noted in 1997. This increase amounted to 3.6 million tonnes. Crude steel production rose by 6.4 million tonnes, an increase of 6.3 per cent, to a new record level of 107.6 million tonnes. China remained the world's largest steel producer. Steel imports fell by 5.1 per cent, while exports rose by 13 per cent.

In 1998, as a result of the restructuring and mergers that are due to take place in the Chinese steel industry, crude steel production will probably decline by around 9 per cent to 98 million tonnes. Exports are expected to fall sharply by around 36.5 per cent due to lower demand in the Asian area. Imports of products with high added value, however, may well increase. Apparent steel consumption should therefore decline by 2.3 per cent in 1998.

In 1999, renewed growth of around 3 per cent in crude steel production will probably boost the volume of production back up over the 100 million tonnes mark. If the course of events in Asia follows the path outlined above, exports could well grow strongly, resulting in a fall in imports with consumption remaining at a level fairly close to that in 1998.

4

TRENDS IN EMPLOYMENT IN THE STEEL INDUSTRY
IN OECD MEMBER COUNTRIES

In 1997, the number of people employed in the steel industry in the OECD area fell by a 26 600, a decline of around 2.9 per cent. Since 1974, the total number of jobs in the steel industry in the OECD area has fallen by 59.4 per cent.

With regard to employment in the steel industry in the European Union, even after completion of the last round of restructuring in 1995, and the heavy job losses which this entailed, efforts to streamline and improve the competitiveness of the industry are continuing. The reduction in the workforce in the EU(15) as a whole amounted to around 4.1 per cent, a loss of 15 700 jobs. This downward trend in employment looks set to continue for several years to come.

In other European countries, although employment appears to have remained stable in Norway and Turkey, the total number of workers employed in the steel industry in Switzerland was down by 18.5 per cent, falling from 1 194 in 1996 to 973 in 1997. The number of workers employed directly in production declined by 37.8 per cent, falling from 897 workers in 1996 to 558 in 1997. This decline is due to restructuring.

Employment in the Japanese steel industry fell by 5.6 per cent, amounting to a loss of some 8 700 jobs. Employment in the steel industry in Korea would appear to have declined by 2.3 per cent.

Employment in the steel industry in Canada declined by merely 0.3 per cent in 1997, bringing the total number employed to 33 403 workers. In 1998, given that the major adjustments in the steel industry have now been completed and that most plants are operating at a high capacity utilisation rate, the level of employment should remain stable at around 33 000 workers.

In the United States, despite increased production the downward trend in employment in the steel industry continued throughout 1997. Total employment fell by 0.7 per cent to 236 000 workers, although the decline in the number of production workers was proportionately less.

◆ Graph 4. **Developments in crude steel capacity and production in the main areas of the world**
Million tonnes

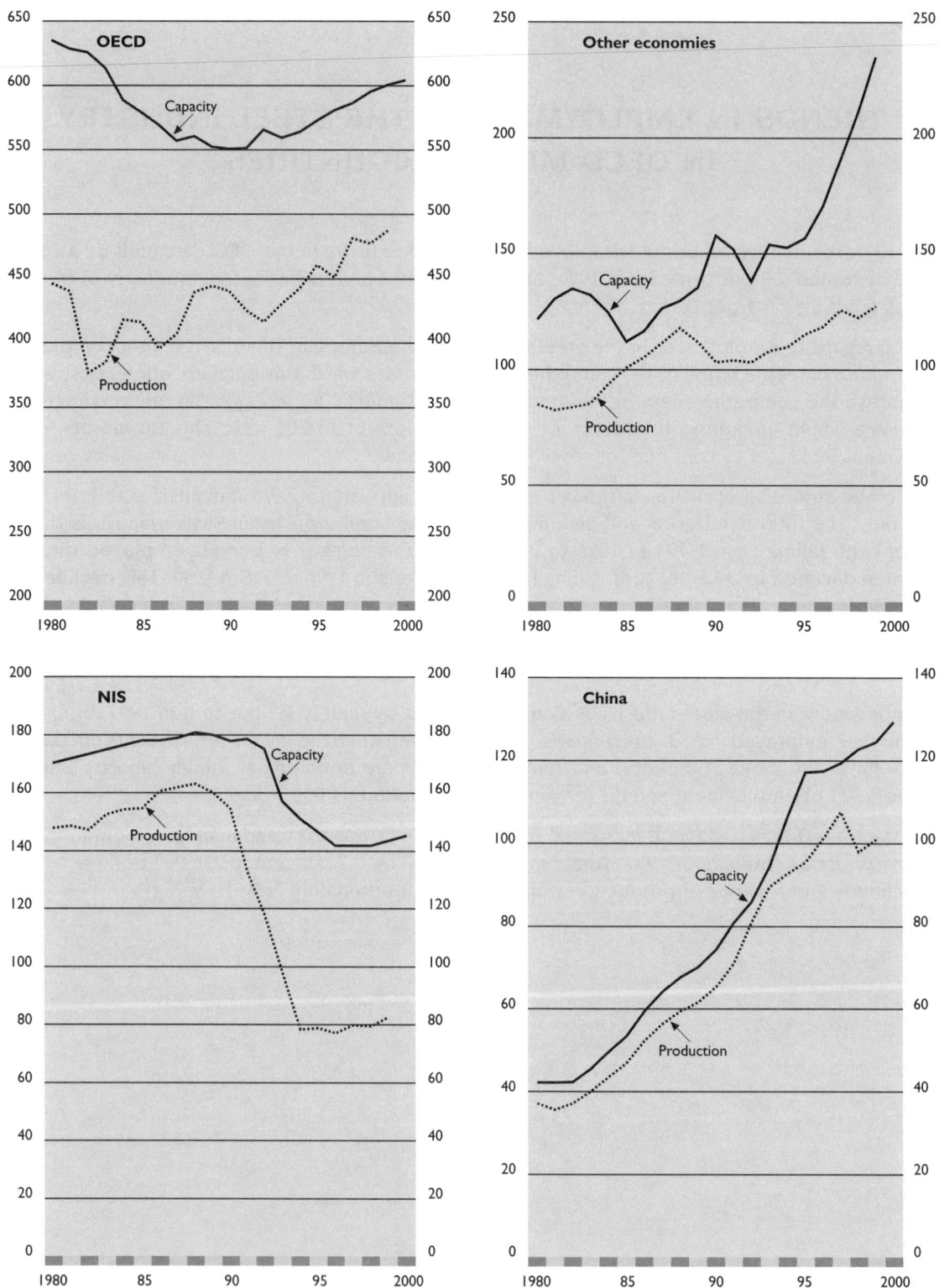

Source: OECD Secretariat.

◆ Graph 5. **Developments in steel consumption and production in the main areas of the world**
Million tonnes finished product equivalent

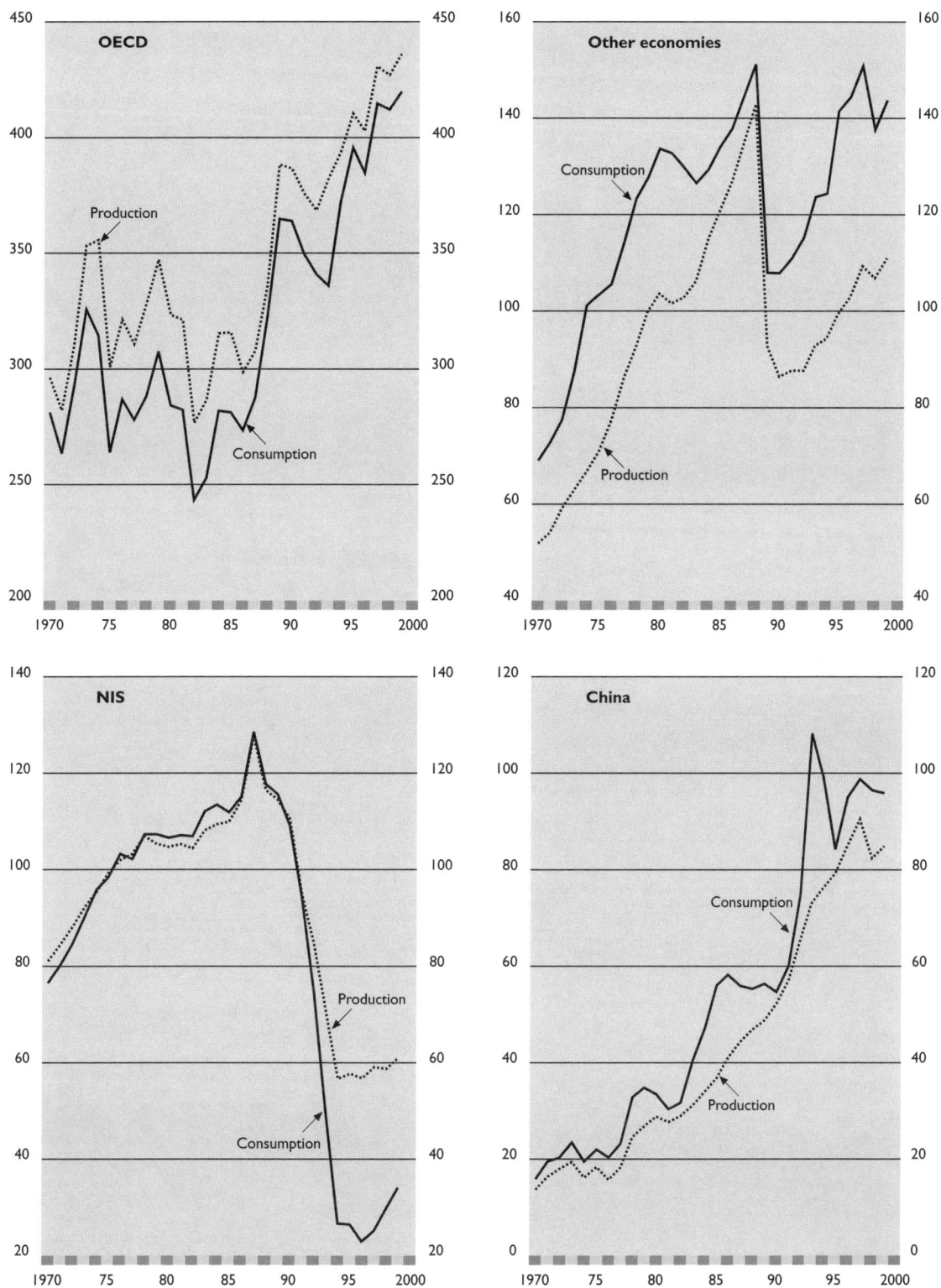

STATISTICAL ANNEX

In order to reflect the expansion of the Organisation which occurred in 1996, data on the Czech Republic, Hungary, Korea and Poland have been included, wherever possible, in the "OECD Total", and to make the tables consistent, the series have been recalculated on this basis for past years.

The main differences with previous reports are the following:

- European Union (15) has replaced EU(12).
- The area "Other Europe" now includes the following countries: Norway, Switzerland, Turkey, the Czech Republic, Hungary and Poland, the last three countries having been removed from the area "Central and Eastern Europe".
- The area "Central and Eastern Europe" now includes only Bulgaria, Romania and the Slovak Republic, as well as (when possible) Albania.
- Brazil, a full participant in the Steel Committee, now appears in almost all tables and has been excluded from the "Other Latin America" area.
- The previous area "Other Asia" has been split in two new areas, "ASEAN(5)" which groups Indonesia, Malaysia, the Philippines, Singapore and Thailand, and a new "Other Asia" which groups the remaining countries, including North Korea which is no longer included with China.
- Wherever possible the Secretariat, in addition to the New Independent States (NIS) area (former Soviet Union) has included the breakdown Russia, Ukraine and other NIS.

Table 1. **Apparent steel consumption** (million tonnes of product equivalent)

Tableau 1. **Consommation apparente d'acier** (en millions de tonnes équivalent produits)

	1990	1993	1994	1995	1996	1997	1998	1999	1997/96 in/en %	1998/97 in/en %	1999/98 in/en %	
United States	86.0	89.9	101.6	97.8	103.8	107.4	108.6	106.5	+3.5	+1.1	-1.9	États-Unis
Canada	9.3	10.8	12.7	13.1	12.6	15.8	14.8	14.7	+26.0	-6.2	-0.7	Canada
EU (15)	119.3	99.5	115.3	129.7	112.2	126.9	132.9	137.1	+12.8	+4.8	+3.1	UE (15)
Other Europe*	25.0	21.5	22.1	25.4	23.8	27.3	26.9	29.2	+14.7	-0.5	+8.6	Autres Europe*
Japan	92.9	76.1	74.9	79.7	78.7	80.9	77.2	79.6	+2.9	-4.6	+3.1	Japon
Australia and New Zealand	5.3	5.6	6.0	6.3	6.1	6.4	6.7	6.6	+4.1	+4.4	-0.6	Australie et Nouvelle-Zélande
Mexico	6.6	7.1	9.1	8.0	10.2	12.1	12.1	12.4	+18.2	0.0	+2.3	Mexique
Korea	19.6	25.2	30.4	35.7	37.3	37.7	32.7	33.8	+1.0	-13.2	+3.3	Corée
OECD	364.0	335.7	372.0	395.6	384.7	414.5	411.9	419.8	+7.8	-0.6	+1.9	OCDE
Brazil	8.6	9.3	11.0	12.1	12.0	14.3	14.7	15.6	+19.2	+2.8	+6.3	Brésil
OECD Steel Committee	372.6	345.0	383.0	407.7	396.7	428.8	426.6	435.4	+8.1	-0.5	+2.1	Comité de l'acier de l'OCDE
Other Latin America	6.8	8.8	9.2	10.8	10.3	11.1	13.5	14.7	+7.7	+21.4	+9.5	Autres Amérique latine
South Africa	4.7	4.2	4.1	4.3	4.4	4.7	4.8	5.0	+6.3	+1.5	+3.6	Afrique du Sud
Other Africa	4.8	7.3	4.7	3.9	4.1	4.0	4.1	4.2	-3.4	+3.8	+2.4	Autres Afrique
Middle East	17.4	22.5	23.2	24.2	26.4	28.7	28.8	29.2	+8.5	+0.5	+1.2	Moyen-Orient
India	12.9	15.1	16.8	19.1	20.5	20.6	19.6	19.6	+0.4	-5.1	+0.4	Inde
ASEAN (5)**	17.3	23.6	26.8	33.3	34.2	31.5	23.1	24.3	-8.1	-26.4	+5.2	ASEAN (5)**
Other Asia	23.4	29.4	24.5	29.5	27.8	31.2	25.1	26.6	+12.9	-19.5	+5.3	Autres Asie
Total	87.3	110.9	109.3	125.1	127.7	131.8	119.0	123.6	+3.2	-9.7	+3.9	Total
Central and Eastern Europe	11.9	3.4	4.1	4.3	4.6	4.7	5.0	5.7	+3.3	+5.9	+13.4	Europe centrale et orientale
of which:												dont :
Romania	6.3	2.1	2.6	2.8	3.1	2.9	3.1	3.4	-7.3	+5.2	+10.5	Roumanie
Slovak Republic	4.3	0.7	0.6	0.6	1.0	1.1	1.1	1.3	+15.5	+1.8	+10.5	République slovaque
NIS Total	109.2	49.7	26.7	26.5	22.9	25.2	29.8	34.1	+9.7	+18.4	+14.5	Total NEI
Russia	..	28.1	14.2	17.6	13.8	13.9	15.6	19.0	+1.0	+12.1	+22.0	Russie
Ukraine	..	16.3	8.7	6.2	6.3	7.8	9.9	10.1	+23.5	+27.0	+2.4	Ukraine
Other NIS	..	5.3	3.8	2.7	2.9	3.5	4.3	5.0	+21.7	+23.6	+15.3	Autres NEI
China	54.8	108.3	99.2	84.3	95.1	98.7	96.5	95.9	+3.9	-2.3	-0.6	Chine
World	635.8	617.3	622.3	647.9	647.0	689.2	676.9	694.7	+6.5	-1.8	+2.6	Monde

* Includes: Norway, Turkey, Iceland, former Yugoslavia, Czech Republic, Hungary and Poland.
** Includes: Indonesia, Malaysia, Philippines, Singapore, Thailand.
Source: OECD Secretariat.

* Comprend : Islande, Norvège, Turquie, ex-Yougoslavie, République tchèque, Hongrie et Pologne.
** Comprend : Indonésie, Malaisie, Philippines, Singapour et Thaïlande.
Source : Secrétariat de l'OCDE.

Table 2. **Trade balance (– = net exports, + = net imports)** (million tonnes)

Tableau 2. **Balance des échanges (– = exportations nettes, + = importations nettes)** (en millions de tonnes)

	1990	1992	1993	1994	1995	1996	1997	1998	1999	
EU (15)	-13.0	-12.5	-29.6	-20.6	-9.8	-19.7	-16.7	-13.6	-11.3	UE (15)
Japan	-9.5	-12.1	-16.6	-16.6	-15.1	-13.3	-16.4	-13.1	-14.3	Japon
Total	-22.5	-24.6	-46.2	-37.2	-24.9	-33.0	-33.1	-26.7	-25.6	Total
United States	11.7	11.6	14.3	23.8	16.1	22.3	23.2	22.6	19.2	États-Unis
Canada	-1.0	-2.4	-1.5	0.8	0.7	-0.1	2.5	1.3	1.2	Canada
Korea	-1.9	-4.6	-5.6	-1.1	1.4	1.0	-2.0	-4.8	-4.5	Corée
Other Europe	-7.3	-6.7	-4.9	-6.2	-3.9	-5.3	-4.6	-5.2	-4.5	Autres Europe
Australia and New Zealand	-1.2	-1.6	-2.2	-2.4	-2.1	-2.2	-2.2	-2.1	-2.0	Australie et Nouvelle-Zélande
Mexico	-0.4	0.8	-0.4	0.7	-2.0	-0.7	0.2	0.2	-0.1	Mexique
Total	-0.1	-2.9	-0.3	15.6	10.2	15.0	17.1	12.0	9.3	Total
OECD	-22.6	-27.4	-46.4	-21.6	-14.7	-18.1	-16.0	-14.9	-16.3	OCDE
Brazil	-8.8	-11.6	-12.0	-10.9	-9.4	-9.9	-8.4	-8.2	-7.8	Brésil
OECD Steel Committee	-31.4	-39.0	-58.4	-32.5	-24.1	-28.0	-24.4	-23.1	-24.1	Comité de l'acier de l'OCDE
Other Latin America	-1.0	1.3	0.7	0.5	1.7	0.4	0.6	2.9	3.3	Autres Amérique latine
South Africa	-2.8	-3.8	-3.5	-3.5	-3.5	-2.7	-2.7	-2.8	-3.1	Afrique du Sud
Other Africa	3.2	6.5	6.0	3.5	2.8	3.2	3.2	3.4	3.2	Autres Afrique
Middle East	11.2	12.5	13.0	13.1	13.8	15.1	16.5	17.5	18.0	Moyen-Orient
India	1.1	0.7	0.1	0.8	0.9	0.7	0.8	0.7	-0.2	Inde
ASEAN (5)	12.2	14.9	16.6	19.8	24.1	24.4	21.9	15.1	15.9	ASEAN (5)
Other Asia	8.1	10.9	18.0	9.6	15.7	16.8	15.3	9.1	9.3	Autres Asie
Total	32.0	43.0	50.9	43.8	55.5	57.9	55.6	45.9	46.4	Total
Central and Eastern Europe	-1.8	-2.9	-5.6	-5.5	-6.3	-5.2	-5.8	-5.8	-5.5	Europe centrale et orientale
Romania	-1.3	-0.6	-2.2	-2.0	-2.4	-1.7	-2.4	-2.4	-2.4	Roumanie
Slovak Republic	-0.2	-1.7	-2.5	-2.6	-2.7	-2.1	-2.1	-2.2	-2.1	République slovaque
NIS	-1.4	-9.9	-21.0	-30.0	-31.2	-33.9	-33.9	-29.0	-27.0	NEI
China	2.5	9.3	35.4	23.1	5.0	9.9	8.3	14.1	11.0	Chine
Unspecified	0.1	0.5	1.3	1.6	1.1	0.7	0.2	2.1	0.8	Non spécifié

Source: OECD Secretariat.

Source : Secrétariat de l'OCDE.

43

Table 3. **Crude steel production** (million tonnes)

Tableau 3. **Production d'acier brut** (en millions de tonnes)

	1990	1993	1994	1995	1996	1997	1998	1999	1997/96 in/en %	1998/97 in/en %	1999/98 in/en %	
United States	89.7	88.8	91.2	95.2	94.7	97.5	99.5	101.0	+3.0	+2.0	+1.5	États-Unis
Canada	12.3	14.4	13.9	14.4	14.7	15.5	15.7	15.7	+4.9	+1.6	0.0	Canada
EU (15)	148.4	144.2	151.7	155.7	146.9	159.9	163.4	165.1	+8.8	+2.1	+1.1	UE (15)
Other Europe	40.7	32.5	34.7	35.7	34.7	37.5	37.7	39.4	+8.2	+0.5	+4.6	Autres Europe
Japan	110.3	99.6	98.3	101.6	98.8	104.6	97.0	100.9	+5.8	-7.2	+4.0	Japon
Australia and New Zealand	7.4	8.6	9.2	9.3	9.2	9.5	9.7	9.6	+2.8	+2.2	-1.5	Australie et Nouvelle-Zélande
Korea	23.1	33.0	33.8	36.8	38.9	42.6	40.2	41.0	+9.4	-5.5	+2.0	Corée
Mexico	8.7	9.2	10.3	12.2	13.2	14.3	14.4	14.8	+8.2	+0.7	+3.2	Mexique
OECD	440.7	430.3	443.0	460.9	451.1	481.2	477.5	487.5	+6.7	-0.8	+2.1	OCDE
Brazil	20.6	25.2	25.7	25.1	25.2	26.2	26.4	27.1	+3.6	+1.0	+2.4	Brésil
OECD Steel Committee	461.3	455.5	468.7	486.0	476.3	507.4	503.9	514.6	+6.5	-0.7	+2.1	Comité de l'acier de l'OCDE
Other Latin America	9.2	9.3	10.0	10.5	11.4	12.0	12.1	13.1	+5.7	+0.6	+8.5	Autres Amérique latine
South Africa	8.6	8.7	8.5	8.7	8.0	8.3	8.5	9.0	+3.9	+2.3	+5.5	Afrique du Sud
Other Africa	2.0	1.6	1.5	1.4	1.2	0.9	0.9	1.3	-20.3	-6.4	+42.0	Autres Afrique
Middle East	6.3	10.6	11.3	11.7	12.7	13.5	12.6	12.5	+7.0	-6.9	-0.8	Moyen-Orient
India	15.0	18.2	19.3	22.0	23.8	23.6	22.4	23.5	0.0	-5.5	+4.6	Inde
ASEAN (5)	5.8	7.7	7.7	10.2	10.9	10.6	8.9	9.4	-2.7	-16.0	+5.0	ASEAN (5)
Other Asia	17.9	15.7	13.4	12.4	13.2	17.1	17.4	17.5	+29.5	+1.8	+0.6	Autres Asie
Total	64.8	71.8	71.7	76.9	81.2	86.0	82.8	86.3	+5.9	-3.7	+4.2	Total
Central and Eastern Europe	17.4	11.3	12.2	13.2	12.1	13.1	13.4	13.9	+7.8	+2.3	+4.0	Europe centrale et orientale
of which:												dont :
Romania	9.8	5.5	5.8	6.6	6.1	6.7	6.9	7.3	+9.7	+2.8	+5.7	Roumanie
Slovak Republic	5.5	3.9	4.0	3.9	3.6	3.8	3.8	3.9	+5.3	+1.3	+1.6	République slovaque
NIS	154.4	98.1	78.3	78.8	77.1	79.8	79.4	82.6	+3.5	-0.4	+4.0	NEI
of which:												dont :
Russia	..	58.4	48.8	51.3	49.2	48.4	46.7	47.9	-1.5	-3.6	+2.7	Russie
Ukraine	..	32.6	24.1	22.3	22.3	24.7	26.0	27.3	+6.0	+5.1	+5.0	Ukraine
Other NIS	..	7.2	5.4	5.2	5.6	6.6	6.7	7.4	+17.7	+2.4	+9.2	Autres NEI
China	65.4	89.5	92.6	95.4	101.2	107.6	98.0	100.9	+6.3	-8.9	+3.0	Chine
World	763.8	726.2	723.5	750.3	747.9	794.1	777.4	798.3	+6.2	-2.1	+2.7	Monde

Source : OECD Secretariat.

Source : Secrétariat de l'OCDE.

Table 4. **Steel production, consumption and trade** (million tonnes)

Tableau 4. **Production, consommation et échanges d'acier** (en millions de tonnes)

1996	Production			Imports/ Importations	Exports/ Exportations	Balance	Apparent consumption/ Consommation apparente	1996
	Crude steel/ Acier brut	Via c.c.	Product eq./ Équiv. produits					
United States	94.7	88.3	81.6	26.9	4.7	22.3	103.8	États-Unis
Canada	14.7	14.4	12.7	4.7	4.8	-0.1	12.6	Canada
EU (15)	146.9	138.4	131.8	12.5	32.2	-19.7	112.2	UE (15)
Other Europe	34.7	24.4	29.1	10.7	16.1	-5.3	23.8	Autres Europe
Japan	98.8	95.2	92.0	5.9	19.3	-13.3	78.7	Japon
Australia and New Zealand	9.2	9.2	8.3	1.4	3.6	-2.2	6.1	Australie et Nouvelle-Zélande
Korea	38.9	38.0	36.3	11.1	10.0	1.0	37.3	Corée
Mexico	13.2	10.5	11.0	0.9	1.6	-0.7	10.2	Mexique
OECD	451.1	418.4	402.8	74.1	92.2	-18.1	384.7	OCDE
Brazil	25.2	18.1	21.9	0.4	10.3	-9.9	12.0	Brésil
OECD Steel Committee	476.3	436.5	424.7	74.5	102.5	-28.0	396.7	Comité de l'acier de l'OCDE
Other Latin America	11.4	9.9	9.9	4.2	3.7	0.4	10.3	Autres Amérique latine
South Africa	8.0	7.3	7.1	0.3	3.0	-2.7	4.4	Afrique du Sud
Other Africa	1.2	0.3	1.0	3.5	0.4	3.2	4.1	Autres Afrique
Middle East	12.7	12.2	11.4	16.0	1.0	15.1	26.4	Moyen-Orient
India	23.8	11.8	19.9	2.1	1.4	0.7	20.5	Inde
ASEAN (5)	10.9	10.6	9.8	27.8	3.3	24.4	34.2	ASEAN (5)
Other Asia	13.2	12.3	12.2	19.3	3.8	15.5	27.7	Autres Asie
Total	81.2	64.4	71.3	73.2	16.6	56.6	127.6	Total
Central and Eastern Europe	12.1	7.3	9.8	1.5	6.7	-5.2	4.6	Europe centrale et orientale
of which:								dont :
Romania	6.1	3.1	4.8	0.6	2.3	-1.7	3.1	Roumanie
Slovak Republic	3.6	3.6	3.1	0.8	2.9	-2.1	1.0	République Slovaque
NIS	77.1	26.8	56.8	7.0	40.9	-33.9	22.9	NEI
of which:								dont :
Russia	49.2	20.1	36.6	4.2	27.0	-22.8	13.8	Russie
Ukraine	22.3	4.6	16.1	0.9	10.7	-9.8	6.3	Ukraine
Other NIS	5.6	2.2	4.2	1.9	3.2	-1.3	2.9	Autres NEI
China	101.2	54.0	85.1	16.4	6.5	9.9	95.1	Chine
World	747.9	588.9	647.6	172.3	173.0	-0.7	647.0	Monde

Source: OECD Secretariat.

Source : Secrétariat de l'OCDE.

Table 5. **Steel production, consumption and trade** (million tonnes)

Tableau 5. **Production, consommation et échanges d'acier** (en millions de tonnes)

1997	Crude steel/ Acier brut	Via c.c.	Product eq./ Équiv. produits	Imports/ Importations	Exports/ Exportations	Balance	Apparent consumption/ Consommation apparente	1997
United States	97.5	92.3	84.2	28.9	5.6	23.21	107.4	États-Unis
Canada	15.5	15.2	13.3	6.7	4.2	2.5	15.8	Canada
EU (15)	159.9	150.9	143.6	13.3	30.0	-16.7	126.9	UE (15)
Other Europe	37.5	29.3	31.9	12.4	17.0	-4.6	27.3	Autres Europe
Japan	104.6	100.8	97.4	6.4	22.8	-16.4	80.9	Japon
Australia and New Zealand	9.5	9.4	8.6	1.5	3.7	-2.2	6.4	Australie et Nouvelle-Zélande
Korea	42.6	41.6	39.7	9.3	11.3	-2.0	37.7	Corée
Mexico	14.3	11.4	11.9	1.1	0.8	0.2	12.1	Mexique
OECD	481.2	450.9	430.5	79.5	95.5	-16.0	414.5	OCDE
Brazil	26.2	18.8	22.6	0.8	9.2	-8.4	14.3	Brésil
OECD Steel Committee	507.4	469.7	453.1	80.3	104.7	-24.4	428.8	Comité de l'acier de l'OCDE
Other Latin America	12.0	10.5	10.5	4.2	3.6	0.6	11.1	Autres Amérique latine
South Africa	8.3	7.6	7.4	0.3	3.0	-2.7	4.7	Afrique du Sud
Other Africa	0.9	0.2	0.8	3.5	0.3	3.2	4.0	Autres Afrique
Middle East	13.5	13.1	12.2	17.5	1.0	16.5	28.7	Moyen-Orient
India	23.8	11.9	19.9	2.2	1.5	0.8	20.6	Inde
ASEAN (5)	10.6	10.3	9.6	24.2	2.3	21.9	31.5	ASEAN (5)
Other Asia	17.2	16.3	15.9	19.8	4.5	15.4	31.3	Autres Asie
Total	86.3	69.9	76.3	71.7	16.2	55.7	131.9	Total
Central and Eastern Europe	13.1	7.8	10.6	1.6	7.4	-5.8	4.7	Europe centrale et orientale
of which:								dont :
Romania	6.7	3.4	5.3	0.6	3.0	-2.4	2.9	Roumanie
Slovak Republic	3.8	3.8	3.3	0.7	2.9	-2.1	1.1	République slovaque
NIS	79.8	30.3	59.1	7.7	41.6	-33.9	25.2	NEI
of which:								dont :
Russia	48.4	22.7	36.4	4.6	27.1	-22.5	13.9	Russie
Ukraine	24.7	5.0	17.8	1.0	11.0	-10.0	7.8	Ukraine
Other NIS	6.6	2.6	4.9	2.1	3.5	-1.4	3.5	Autres NEI
China	107.6	57.5	90.5	15.5	7.3	8.3	98.7	Chine
World	794.1	635.0	689.4	176.8	177.0	-0.2	689.2	Monde

Source: OECD Secretariat.

Source : Secrétariat de l'OCDE.

Table 6. **Steel production, consumption and trade** (million tonnes)

Tableau 6. **Production, consommation et échanges d'acier** (en millions de tonnes)

1998	Production			Imports/ Importations	Exports/ Exportations	Balance	Apparent consumption/ Consommation apparente	1998
	Crude steel/ Acier brut	Via c.c.	Product eq./ Équiv. produits					
United States	99.5	95.1	86.0	28.0	5.4	22.6	108.6	États-Unis
Canada	15.7	15.4	13.5	5.5	4.2	1.3	14.8	Canada
EU (15)	163.4	153.7	146.6	14.3	28.0	–13.6	132.9	UE (15)
Other Europe	37.7	29.7	32.1	12.1	17.3	–5.2	26.9	Autres Europe
Japan	97.0	93.5	90.3	5.1	18.2	–13.1	77.2	Japon
Australia and New Zealand	9.7	9.7	8.8	1.5	3.6	–2.1	6.7	Australie et Nouvelle-Zélande
Korea	40.2	39.3	37.5	7.6	12.4	–4.8	32.7	Corée
Mexico	14.4	11.5	11.9	1.0	0.8	0.2	12.1	Mexique
OECD	477.5	447.9	426.8	75.0	89.9	–14.9	411.9	OCDE
Brazil	26.4	19.0	22.9	0.3	8.5	–8.2	14.7	Brésil
OECD Steel Commitee	503.9	466.9	449.7	75.3	98.4	–23.1	426.6	Comité de l'acier de l'OCDE
Other Latin America	12.1	10.8	10.5	6.0	3.1	2.9	13.5	Autres Amérique latine
South Africa	8.5	7.7	7.6	0.2	3.0	–2.8	4.8	Afrique du Sud
Other Africa	0.9	0.2	0.7	3.6	0.2	3.4	4.1	Autres Afrique
Middle East	12.6	12.2	11.3	18.6	1.1	17.5	28.8	Moyen-Orient
India	22.4	12.3	18.9	2.6	1.9	0.7	19.6	Inde
ASEAN (5)	8.9	8.7	8.0	17.1	2.0	15.1	23.1	ASEAN (5)
Other Asia	17.3	16.4	16.0	12.4	3.2	9.2	25.2	Autres Asie
Total	82.7	68.3	73	60.5	14.5	46	119.1	Total
Central and Eastern Europe	13.4	8.0	10.8	1.7	7.5	–5.8	5.0	Europe centrale et orientale
of which:								dont :
Romania	6.9	3.5	5.5	0.7	3.1	–2.4	3.1	Roumanie
Slovak Republic	3.8	3.8	3.3	0.8	2.9	–2.2	1.1	République slovaque
NIS	79.4	29.8	58.8	7.6	36.6	–29.0	29.8	NEI
of which:								dont :
Russia	46.7	21.9	35.1	4.1	23.6	–19.5	15.6	Russie
Ukraine	26.0	5.3	18.7	1.2	10.0	–8.8	9.9	Ukraine
Other NIS	6.7	2.6	5.0	2.3	3.0	–0.7	4.3	Autres NEI
China	98.0	52.4	82.4	18.7	4.6	14.1	97.0	Chine
World	777.4	625.3	674.8	163.6	161.6	2.0	676.8	Monde

Source: OECD Secretariat.

Source : Secrétariat de l'OCDE.

Table 7. **Steel production, consumption and trade** (million tonnes)

Tableau 7. **Production, consommation et échanges d'acier** (en millions de tonnes)

1999	Production Crude steel/ Acier brut	Production Via c.c.	Production Product eq./ Équiv. produits	Imports/ Importations	Exports/ Exportations	Balance	Apparent consumption/ Consommation apparente	1999
United States	101.0	96.6	87.3	23.2	4.0	19.2	106.5	États-Unis
Canada	15.7	15.4	13.5	5.3	4.1	1.2	14.7	Canada
EU (15)	165.1	156.7	148.3	14.5	25.8	-11.3	137.1	UE (15)
Other Europe	39.4	31.9	3.7	11.7	16.2	-4.5	29.2	Autres Europe
Japan	100.9	97.2	93.9	5.0	19.3	-14.3	79.6	Japon
Australia and New Zealand	9.6	9.5	8.6	1.7	3.7	-2.0	6.6	Australie et Nouvelle-Zélande
Korea	41.0	40.1	38.3	5.5	10.0	-4.5	33.8	Corée
Mexico	14.8	12.6	12.4	1.0	1.1	-0.0	12.4	Mexique
OECD	487.5	460.7	436.1	67.9	84.2	-16.3	419.8	OCDE
Brazil	27.1	19.4	23.4	0.6	8.4	-7.8	15.6	Brésil
OECD Steel Commitee	514.6	480.1	459.5	68.5	92.6	-24.1	435.4	Comité de l'acier de l'OCDE
Other Latin America	13.1	11.7	11.4	6.2	2.9	3.3	14.7	Autres Amérique latine
South Africa	9.0	8.2	8.0	0.3	3.3	-3.1	5.0	Afrique du Sud
Other Africa	1.3	0.3	1.0	3.4	0.2	3.2	4.2	Autres Afrique
Middle East	12.5	12.1	11.2	18.8	0.9	18.0	29.2	Moyen-Orient
India	23.5	12.9	19.8	2.0	2.2	-0.2	19.6	Inde
ASEAN (5)	9.4	9.1	8.4	18.4	2.5	15.9	24.3	ASEAN (5)
Other Asia	17.7	16.8	16.4	14.6	4.5	10.1	26.5	Autres Asie
Total	86.5	71.1	76.2	63.7	16.5	47.2	123.5	Total
Central and Eastern Europe	13.9	8.2	11.2	1.8	7.3	-5.5	5.7	Europe centrale et orientale
of which:								dont :
Romania	7.3	3.7	5.8	0.7	3.1	-2.4	3.4	Roumanie
Slovak Republic	3.9	3.9	3.4	0.7	2.8	-2.1	1.3	République slovaque
NIS	82.6	30.8	61.1	7.8	34.8	-27.0	34.1	NEI
of which:								dont :
Russia	47.9	22.4	36.0	4.5	21.5	-17.0	19.0	Russie
Ukraine	27.3	5.5	19.6	1.0	10.5	-9.5	10.1	Ukraine
Other NIS	7.4	2.9	5.5	2.3	2.8	-0.5	5.0	Autres NEI
China	100.9	54.0	84.9	16.5	5.5	11.0	95.9	Chine
World	798.3	643.7	693.1	158.1	156.7	1.4	694.5	Monde

Source: OECD Secretariat.

Source : Secrétariat de l'OCDE.

Table 8. **The steel markets in the United States, the European Union and Japan**

Tableau 8. **Les marchés de l'acier aux États-Unis, dans l'Union européenne et au Japon**

	United States/États-Unis				EU (15)/UE (15)				Japan/Japon				
	1996	1997	1998	1999	1996	1997	1998	1999	1996	1997	1998	1999	
	In million product tonnes/En millions de tonnes produit												
Real consumption	101.5	107.2	108.8	107.5	120.1	122.9	132.9	136.0	79.0	80.3	76.4	80.1	Consommation réelle
Stocks, consumers and merchants	+0.8	+0.1	-0.2	-0.8	-4.0	+5.0	0.0	+0.5	-0.3	+0.6	+0.8	-0.5	Stocks des consom. et des marchands
Market	102.3	107.3	108.6	106.7	116.1	127.9	132.9	136.5	78.7	80.9	77.2	79.6	Marché
Imports	26.9	28.8	27.8	23.2	12.5	13.3	14.3	14.5	5.9	6.4	5.1	5.0	Importations
Exports	4.7	5.6	5.4	4.0	32.2	30.0	28.0	25.8	19.2	22.8	18.2	19.3	Exportations
Deliveries	80.4	84.1	86.2	87.5	135.8	144.6	146.6	147.8	92.0	97.4	90.3	93.9	Livraisons
Producers' stocks	+1.2	+0.1	-0.2	-0.2	-4.0	-1.0	0.0	+0.5	0	0	0	0	Stocks des prod
Production	81.6	84.2	86.0	87.3	131.8	143.6	146.6	148.3	92.0	97.4	90.3	93.9	Production
	In million tonnes of crude steel/En millions de tonnes d'acier brut												
Crude steel production	94.7	97.5	99.5	101.0	146.9	159.9	163.4	165.1	98.8	104.6	97.0	100.9	Production d'acier brut
Capacity	105.2	109.3	113.3	113.3	198.8	196.6	197.4	195.1	149.6	149.5	149.5	149.5	Capacité
	In %/En %												
Capacity utilisation	90.0	89.2	87.8	89.1	73.9	81.3	82.8	84.6	66.0	70.0	64.9	67.5	Utilisation de capacité
Import share	26.3	26.8	25.6	21.7	10.8	10.4	10.8	10.6	7.5	7.9	6.6	6.3	Part d'importation

Source: OECD Secretariat.

Source : Secrétariat de l'OCDE.

Table 9. **Steel markets in EU countries** (million tonnes of product equivalent)

Tableau 9. **Les marchés de l'acier dans les pays de l'UE** (en millions de tonnes équivalent produits)

Germany/Allemagne

	1996	1997	1998	1999	
Real consumption	32.1	30.1	31.3	29.2	Consommation réelle
Stocks	-0.5	-0.1	-0.1	+0.1	Stocks
Apparent consumption	31.5	30.0	31.2	29.3	Consommation apparente
Imports	15.9	15.3	16.5	17.2	Importations
Exports	20.0	25.7	25.7	27.0	Exportations
Production	35.7	40.4	40.4	39.1	Production
Crude steel production	39.8	45.0	45.0	43.5	Production d'acier brut
Capacity	50.7	51.6	51.6	51.6	Capacité
Capacity utilisation in %	79	87	87	84	Utilisation de la capacité en %

France

	1996	1997	1998	1999	
Real consumption	14.8	15.0	16.5	17.7	Consommation réelle
Stocks	-0.9	+0.5	+0.2	-0.1	Stocks
Apparent consumption	13.9	15.5	16.7	17.6	Consommation apparente
Imports	11.1	11.9	11.0	11.0	Importations
Exports	13.1	14.1	13.1	12.6	Exportations
Production	15.8	17.7	18.8	19.2	Production
Crude steel production	17.6	19.8	21.0	21.4	Production d'acier brut
Capacity	22.0	24.2	24.3	24.3	Capacité
Capacity utilisation in %	80	82	86	88	Utilisation de la capacité en %

Italy/Italie

	1996	1997	1998	1999	
Real consumption	21.4	25.6	28.6	29.9	Consommation réelle
Stocks	-1.4	+1.0	-0.2	-0.1	Stocks
Apparent consumption	20.0	26.6	28.4	29.8	Consommation apparente
Imports	11.2	13.0	13.0	12.9	Importations
Exports	13.0	9.6	9.3	9.0	Exportations
Production	21.8	23.2	24.6	25.9	Production
Crude steel production	24.3	25.8	27.4	28.8	Production d'acier brut
Capacity	41.8	35.5	35.6	35.6	Capacité
Capacity utilisation in %	58	73	77	81	Utilisation de la capacité en %

United Kingdom/Royaume-Uni

	1996	1997	1998	1999	
Real consumption	13.0	13.5	13.6	14.0	Consommation réelle
Stocks	-0.1	+0.4	+0.1	+0.2	Stocks
Apparent consumption	12.9	13.9	13.7	14.2	Consommation apparente
Imports	6.1	5.8	5.4	6.0	Importations
Exports	9.2	8.4	7.8	7.8	Exportations
Production	16.0	16.5	16.5	16.0	Production
Crude steel production	18.0	18.5	18.1	17.9	Production d'acier brut
Capacity	20.7	20.3	20.3	20.3	Capacité
Capacity utilisation in %	87	91	89	88	Utilisation de la capacité en %

Netherlands/Pays-Bas

	1996	1997	1998	1999	
Real consumption	4.9	5.1	5.0	5.3	Consommation réelle
Stocks	-0.2	0.0	+0.1	+0.2	Stocks
Apparent consumption	4.7	5.0	5.1	5.5	Consommation apparente
Imports	5.5	5.8	5.6	5.8	Importations
Exports	6.5	6.8	6.4	6.3	Exportations
Production	5.7	6.0	6.0	6.0	Production
Crude steel production	6.3	6.6	6.7	6.7	Production d'acier brut
Capacity	6.8	6.8	6.8	6.8	Capacité
Capacity utilisation in %	93	91	99	99	Utilisation de la capacité en %

Belgium and Luxembourg/Belgique et Luxembourg

	1996	1997	1998	1999	
Real consumption	3.8	6.2	7.2	8.0	Consommation réelle
Stocks	-0.3	+0.2	+0.2	+0.1	Stocks
Apparent consumption	3.5	6.4	7.4	8.1	Consommation apparente
Imports	6.2	7.1	7.5	8.0	Importations
Exports	14.5	12.5	12.2	12.0	Exportations
Production	11.9	11.9	12.0	12.1	Production
Crude steel production	13.3	13.4	13.5	13.6	Production d'acier brut
Capacity	18.8	18.2	18.3	18.4	Capacité
Capacity utilisation in %	71	74	74	74	Utilisation de la capacité en %

Spain/Espagne

	1996	1997	1998	1999	
Real consumption	11.6	12.9	14.0	15.5	Consommation réelle
Stocks	-0.6	+0.6	+0.1	+0.1	Stocks
Apparent consumption	11.0	13.5	14.1	15.6	Consommation apparente
Imports	5.5	6.7	6.4	6.3	Importations
Exports	5.5	5.5	5.1	4.7	Exportations
Production	10.9	12.4	12.8	14.0	Production
Crude steel production	12.2	13.8	14.2	15.5	Production d'acier brut
Capacity	17.7	18.2	18.6	18.6	Capacité
Capacity utilisation in %	69	74	76	83	Utilisation de la capacité en %

Finland/Finlande

	1996	1997	1998	1999	
Real consumption	1.9	2.4	2.8	2.8	Consommation réelle
Stocks	-0.1	+0.2	+0.1	+0.1	Stocks
Apparent consumption	1.8	2.6	2.9	2.9	Consommation apparente
Imports	1.3	1.3	1.2	1.2	Importations
Exports	2.6	2.1	2.0	2.0	Exportations
Production	3.1	3.5	3.7	3.7	Production
Crude steel production	3.3	3.7	3.9	4.0	Production d'acier brut
Capacity	4.2	4.2	4.3	4.3	Capacité
Capacity utilisation in %	79	88	91	92	Utilisation de la capacité en %

Sweden/Suède

	1996	1997	1998	1999	
Real consumption	3.9	3.4	3.4	3.6	Consommation réelle
Stocks	-0.3	+0.2	-0.2	+0.1	Stocks
Apparent consumption	3.6	3.6	3.4	3.7	Consommation apparente
Imports	2.8	2.8	2.8	2.8	Importations
Exports	3.6	3.9	4.0	3.8	Exportations
Production	4.4	4.6	4.7	4.8	Production
Crude steel production	4.9	5.2	5.3	5.3	Production d'acier brut
Capacity	5.5	6.2	6.4	6.5	Capacité
Capacity utilisation in %	89	84	83	82	Utilisation de la capacité en %

Austria/Autriche

	1996	1997	1998	1999	
Real consumption	3.0	3.0	3.4	3.3	Consommation réelle
Stocks	-0.2	+0.2	-0.1	+0.1	Stocks
Apparent consumption	2.8	3.2	3.3	3.4	Consommation apparente
Imports	1.8	2.1	2.1	2.1	Importations
Exports	3.1	3.6	3.5	3.4	Exportations
Production	4.0	4.7	4.7	4.7	Production
Crude steel production	4.4	5.2	5.2	5.2	Production d'acier brut
Capacity	4.6	5.6	5.6	5.6	Capacité
Capacity utilisation in %	97	93	93	93	Utilisation de la capacité en %

Rest EU (15)/Reste UE (15)

	1996	1997	1998	1999	
Real consumption	6.5	6.9	7.6	7.6	Consommation réelle
Stocks	-0.2	+0.4	-0.1	+0.2	Stocks
Apparent consumption	6.3	7.3	7.5	7.8	Consommation apparente
Imports	6.1	6.8	6.9	6.9	Importations
Exports	2.3	2.2	2.2	1.9	Exportations
Production	2.5	2.8	2.8	2.8	Production
Crude steel production	2.8	3.0	3.1	3.1	Production d'acier brut
Capacity	6.0	6.1	6.5	6.0	Capacité
Capacity utilisation in %	47	50	48	52	Utilisation de la capacité en %

Note. Trade figures for individual EU countries represent the sum of intra-EU trade and trade with third countries.
Source: OECD Secretariat.

Note : Les chiffres d'échanges pour les pays individuels des UE représentent la somme des échanges avec les pays tiers et des échanges intra-communautaires.
Source : Secrétariat de l'OCDE.

Table 10. **Steel markets in "other Western Europe" and in Mexico** (million tonnes product equivalent)

Tableau 10. **Les marchés de l'acier dans les autres pays d'Europe occidentale et le Mexique** (en millions de tonnes équivalent produits)

	Turkey/Turquie				Iceland and ex-Yugoslavia/ Islande et ex-Yougoslavie				Norway/Norvège				Switzerland/Suisse				
	1996	1997	1998	1999	1996	1997	1998	1999	1996	1997	1998	1999	1996	1997	1998	1999	
Apparent consumption	9.5	10.9	11.2	11.8	1.3	1.3	1.6	2.0	1.3	1.4	1.3	1.4	1.8	2.1	2.1	2.3	Consommation apparente
Imports	3.5	3.8	3.2	2.7	0.9	0.9	1.1	1.2	1.6	1.7	1.6	1.6	1.8	2.1	2.0	2.0	Importations
Exports	5.9	5.7	6.0	5.0	0.5	0.9	1.0	1.1	0.7	0.8	0.8	0.8	0.8	1.0	1.0	0.9	Exportations
Production	12.0	12.8	14.0	14.1	1.0	1.4	1.5	1.9	0.5	0.5	0.6	0.6	0.8	1.1	1.1	1.1	Production
Crude steel production	13.4	14.3	15.6	15.7	1.1	1.6	1.7	2.1	0.5	0.6	0.6	0.6	0.9	1.1	1.2	1.2	Production d'acier brut
Capacity	19.3	20.1	20.1	20.1	2.5	2.5	2.8	2.8	0.6	0.6	0.6	0.6	1.1	1.1	1.1	1.1	Capacité
Capacity utilisation in %	69	71	78	78	44	64	61	75	91	100	100	100	81	100	109	110	Utilisation de capacité en %

	Czech Republic/ République tchèque				Hungary/Hongrie				Poland/Pologne				Mexico/Mexique				
	1996	1997	1998	1999	1996	1997	1998	1999	1996	1997	1998	1999	1996	1997	1998	1999	
Apparent consumption	3.1	3.5	3.7	3.7	1.2	1.4	1.6	1.8	5.6	6.6	6.9	7.2	10.2	12.1	12.1	12.4	Consommation apparente
Imports	1.4	1.5	1.6	1.5	0.6	0.9	0.9	1.0	1.2	1.5	1.7	1.7	0.9	1.1	1.0	1.0	Importations
Exports	3.4	3.5	3.4	3.4	0.9	1.0	0.9	0.9	3.8	4.2	4.2	4.3	1.6	0.8	0.8	1.1	Exportations
Production	5.2	5.5	5.5	5.6	1.6	1.4	1.6	1.7	8.2	9.4	9.4	9.8	11.0	11.9	11.9	12.4	Production
Crude steel production	6.5	6.8	6.7	6.8	1.9	1.7	1.9	2.0	10.4	11.6	11.7	12.2	13.2	14.3	14.4	14.8	Production d'acier brut
Capacity	8.8	8.8	8.8	8.8	1.9	1.9	1.9	1.9	11.7	12.5	12.3	12.4	15.2	15.3	16.5	17.0	Capacité
Capacity utilisation in %	74	77	76	77	100	92	100	105	89	93	95	98	86	93	87	87	Utilisation de capacité en %

Source: OECD Secretariat.

Source : Secrétariat de l'OCDE.

Table 11. **Manpower**

Tableau 11. **Main-d'œuvre**

	Average numbers employed ('000)/ Moyenne des effectifs ('000)							% change/variation		
	1974	1984	1993	1994	1995	1996	1997	1997/1996	1997/1974	
Belgium/Luxembourg	86.6*	51.4	32.2	30.9	29.8	28.4	25.3	−10.9	−70.8	Belgique/Luxembourg
Denmark, Ireland	3.5	2.3	1.8	1.6	1.5	1.5	1.6	+6.7	−54.3	Danemark, Irlande
France	155.7	87.1	41.2	40.4	39.3	38.5	38.3	−0.5	−75.4	France
Germany	230.6	156.5	119.0	100.0	92.5	85.9	82.3	−4.2	−64.3	Allemagne
Greece	8.7	4.2	3.0	2.7	2.5	2.3	2.1	−8.7	−75.9	Grèce
Italy	93.8	81.7	50.4	45.5	42.1	39.2	36.9	−5.9	−60.7	Italie
Netherlands	23.8	18.7	14.6	13.1	12.6	12.3	12.1	−1.6	49.2	Pays-Bas
Portugal	5.0	6.7	3.2	2.9	2.7	2.4	2.1	−12.5	−58.0	Portugal
Spain	89.4	69.2	30.1	26.7	25.3	23.7	23.2	−2.1	−74.0	Espagne
United Kingdom	197.7	62.3	40.2	38.5	37.9	37.0	35.9	−3.0	−81.8	Royaume-Uni
EU (12)	894.8	540.1	335.4	302.3	286.2	271.2	259.5	−4.3	−71.0	UE (12)
Austria	43.0	34.9	16.2	15.4	13.3	12.9	12.4	−3.9	−71.2	Autriche
Finland	8.1	9.0	8.7	8.8	7.2	7.4	7.3	−1.4	−9.9	Finlande
Sweden	51.0	32.2	20.9	20.7	14.5	14.0	13.8	−1.4	−72.9	Suède
EU (15)	996.9	616.2	381.2	347.2	321.2	305.5	293.0	−4.1	−70.6	UE (15)
Norway	7.3*	4.0	1.4	1.3	1.3*	1.3*	1.3*	0	−82.2	Norvège
Switzerland	5.2	3.0	1.9	1.6	1.3	1.2	1.0	−16.7	−80.8	Suisse
Turkey	36.1	35.0	35.2	32.4	29.9*	32.0*	32.0*	0	−11.4	Turquie
Canada	52.2	51.5	33.4	31.5	33.7	33.5	33.4	−0.3	−36.0	Canada
United States	609.5	267.4	238.8	233.5	240.7	237.6	236.0	−0.7	−61.3	États-Unis
Australia	43.2	30.5	26.3*	26.0*	26.0*	26.0*	24.0*	−7.7	−44.4	Australie
Japan	323.9	264.8	193.0	182.7	168.8	155.1	146.4	−5.6	−54.8	Japon
Mexico	45.5	77.6*	56.8	57.0	48.6	55.0*	55.0*	0	+20.9	Mexique
Korea	62.9*	62.9	66.2	59.8*	66.3*	66.5*	65.0*	−2.3	+3.3	Corée
Total OECD	2 183.0*	1 412.9*	1 034.5	1 013.4	937.8*	913.7*	887.1*	−2.9	−59.4	Total OCDE
Czech Republic	..	..			78.6					République tchèque
Hungary	..	..	18.8	16.9	16.0	14.6				Hongrie
Poland	..	..	111.9	95.9	100.2	99.6	91.2*	−8.4		Pologne

Source: OECD Secretariat estimate.

Source : Estimation du Secrétariat de l'OCDE.

OECD PUBLICATIONS, 2, rue André-Pascal, 75775 PARIS CEDEX 16
PRINTED IN FRANCE
(58 98 02 1 P) ISBN 92-64-16190-2 – No. 50369 1998